The weather of the 1780s over Europe

JOHN KINGTON
Climatic Research Unit, University of East Anglia

CAMBRIDGE UNIVERSITY PRESS
Cambridge
New York New Rochelle Melbourne Sydney

CAMBRIDGE UNIVERSITY PRESS
Cambridge, New York, Melbourne, Madrid, Cape Town, Singapore, São Paulo, Delhi

Cambridge University Press
The Edinburgh Building, Cambridge CB2 8RU, UK

Published in the United States of America by Cambridge University Press, New York

www.cambridge.org
Information on this title: www.cambridge.org/9780521113076

First published 1988
This digitally printed version 2009

A catalogue record for this publication is available from the British Library

Library of Congress Cataloguing in Publication data

Kington, John, 1930–
The weather of the 1780s over Europe.
Bibliography
Includes index.
1. Europe–Climate. I. Title.
QC989.A1K56 1988 551.694 87–32739

ISBN 978-0-521-26079-4 hardback
ISBN 978-0-521-11307-6 paperback

The weather of the 1780s over Europe

Contents

We have an almost incalculable series of weather observations, most of which, like a buried treasure, are without any use for science, for no one will take the trouble – to be sure, very great trouble – to derive from the thousands of observations appropriate comparisons and thus to test whether we can get results from them.
Heinrich Brandes

	List of tables and figures	vii
	Acknowledgements	ix
1	Introduction to the project – mapping the weather of the past	1
2	The development of observing during the historical-instrumental period of meteorology (1600–1850)	3
3	Missing links	18
4	A bi-centenary exercise	21
5	Daily synoptic weather maps, 1781–5	27
6	Weather types and circulation patterns of the 1780s	148
	Bibliography and references	165
	Index	166

Tables and Figures

TABLES

2.1 Catalogue of meteorological stations making weather observations in the 1780s. 6
2.2 *Societas Meteorologica Palatina* specifications and plotting symbols for wind force together with Beaufort scale equivalents. 14
2.3 Terms used to describe state of the sky, weather and visibility in ships' logbooks during the eighteenth century. 16
4.1 *Societas Meteorologica Palatina* definitions and plotting symbols for the state of the sky. 24
4.2 *Societas Meteorologica Palatina* definitions and plotting symbols for significant weather. 24
4.3 *Societas Meteorologica Palatina* definitions and plotting symbols for optical and atmospheric phenomena. 24
6.1 Lamb British Isles weather types for May 1783. 149
6.2 Example of the weighted scoring system used with the Lamb British Isles weather types for May 1783. 149
6.3 Catalogue of Lamb circulation types over the British Isles for 1781. 150
6.4 Catalogue of Lamb circulation types over the British Isles for 1782. 151
6.5 Catalogue of Lamb circulation types over the British Isles for 1783. 152
6.6 Catalogue of Lamb circulation types over the British Isles for 1784. 153
6.7 Catalogue of Lamb circulation types over the British Isles for 1785. 154
6.8 Lamb British Isles weather types: monthly and annual frequencies for 1781. 155
6.9 Lamb British Isles weather types: monthly and annual frequencies for 1782. 156
6.10 Lamb British Isles weather types: monthly and annual frequencies for 1783. 156
6.11 Lamb British Isles weather types: monthly and annual frequencies for 1784. 156
6.12 Lamb British Isles weather types: monthly and annual frequencies for 1785. 156
6.13 Lamb British Isles weather types: annual and period average frequencies, 1781–5; long-period average frequencies and extremes, 1861–1980. 157
6.14 Lamb British Isles weather types: period average frequencies for 1900–54; 1781–5; and 1976–80. 158
6.15 Catalogue of *Grosswetterlagen* for 1781. 159
6.16 Catalogue of *Grosswetterlagen* for 1782. 160
6.17 Catalogue of *Grosswetterlagen* for 1783. 161
6.18 Catalogue of *Grosswetterlagen* for 1784. 162
6.19 Catalogue of *Grosswetterlagen* for 1785. 163
6.20 *Grosswetterlagen:* annual and period average frequencies of the three main circulation patterns, 1781–5 and 1881–1980. 163

FIGURES

2.1 Example from the Royal Society scheme for making weather observations as suggested by Robert Hooke. 4
2.2 Meteorological instruments designed by Robert Hooke for the Royal Society scheme of weather observations. 4
2.3 Register of daily observations for October 1781 kept at Dijon, Burgundy by Dr Maret for the *Société Royale de Médecine.* 6
2.4 Mercury barometer and mercury thermometer of the *Societas Meteorologica Palatina.* 12
2.5 Extract from the *Ephemerides* of the *Societas Meteorologica Palatina*, showing daily observations made at Mannheim by Johann Hemmer from 1–15 January, 1785. 13
2.6 Extract from the meteorological register kept at Lyndon Hall, Rutland by Thomas Barker for June and July, 1783. 15
2.7 Extract from the meteorological register kept at Lambhús, Iceland by Rasmús Lievog in May 1782. 16
2.8 Extract from the logbook of H. M. Cutter *Cockatrice* from Saturday, 27 December 1783 to Thursday, 8 January 1784 whilst cruising off the coast of Sussex. 17
3.1 The first synoptic presentation of meteorological observations in relation to the pressure distribution, as devised by Heinrich Brandes in 1820. 19
4.1 Map of stations showing the synoptic coverage available for the 1780s. 23
4.2 Examples of station plots. 25
6.1 *P* index: graph of five-year running means, 1861–1978. 155
6.2 *P* index: graphs of the seasonal pattern of progression for the periods 1861–1978 and 1781–5. 157
6.3 *Grosswetterlagen:* graphs of blocking highs (A+E) and annual mean sunspot numbers, 1780–6. 158

Acknowledgements

Many people and organisations have given their assistance and advice with this project. The author is grateful to all of them and in particular wishes to mention: Professor W. G. V. Balchin, Mrs Jacqueline Church and Professor J. Oliver at the University College of Swansea; Mrs Joyce Cowlard at the National Meteorological Library, Bracknell; Dr J.-P. Desaive and Professor E. le Roy Ladurie at the University of Paris; Mr F. E. Dixon of the Irish Meteorological Service, Dublin; Dr C. Dumstrei and Dr K. Frydendahl of the Danish Meteorological Institute, Copenhagen; Dr G. Farmer, Dr P. D. Jones and Dr P. M. Kelly of the Climatic Research Unit, Norwich; Dr G. M. Helgason and Dr Sjöfn Kristjánsdóttir of the National Library of Iceland, Reykjavik; Miss Jean Kennedy at the Norfolk and Norwich Record Office, Norwich; Professor G. Manley; Dr H. O. Mertins and Dr D. Stranz at the Seewetteramt, Deutscher Wetterdienst, Hamburg; Professor J. Michalczewski at the Institute of Meteorology and Water Management, Warsaw; Dr G. Nicole-Genty at the National Academy of Medicine, Paris; Dr W. Odelberg of the Royal Swedish Academy of Sciences, Stockholm; Mr M. G. Pearson at the University of Edinburgh; Dr C. Pfister at the University of Bern; Miss Daphne Pipe at the National Maritime Museum, Greenwich; Professor M. Puigcerver and Dr J. Viñas Riera at the University of Barcelona; Dr H. Rohde of the Bundesanstalt für Wasserbau, Hamburg; Dr D. J. Schove at St David's College, Beckenham; and Dr H. ten Kate at the Royal Dutch Meteorological Institute, De Bilt.

I would also like to thank the following institutions which have kindly supplied historical weather data for this project:
Bibliothèque Publique de Dijon; British Museum,

London; Bruce Castle, Haringey; Finnish Meteorological Service, Helsinki; Hertford Record Office, Hertford; India Office Library and Records, London; Instituto Municipal de Historia, Barcelona; Royal Norwegian Scientific Society, Trondheim; Météorologie Nationale, Paris; UK Meteorological Office, Bracknell; National Library of Scotland, Edinburgh; Norwegian Meteorological Institute, Oslo; Public Record Office, London; Royal Academy of Sciences and Arts, Barcelona; Royal Horticultural Society, London; Royal Meteorological Society, Bracknell; Royal Society, London; Royal Society of Edinburgh; Scottish Record Office, Edinburgh; Strathclyde Regional Archives, Glasgow; and the Swedish Meteorological and Hydrological Institute, Stockholm.

The author is grateful to Professor H. H. Lamb and Dr H. T. Mörth at the Climatic Research Unit together with Dr Teich and his colleagues at the Deutscher Wetterdienst, Offenbach am Main for helpful discussions and correspondence during the classification of British Isles weather types and *Grosswetterlagen.*

This research has been supported by grants and funds from the UK Meteorological Office, the Wolfson Foundation, the Climatic Research Unit and the University of East Anglia, all of which are gratefully acknowledged.

My special thanks to Dr Tom Wigley, Director of the Climatic Research Unit, for his helpful suggestions with the research and for kindly providing the facilities which have allowed this project to be continued. Mr David Mew of the School of Environmental Sciences in the University of East Anglia and Mrs Susan Vine are thanked for their assistance in the preparation of the charts for publication, as also is Mr Stuart Robinson of the School of Environmental Sciences for photographing the illustrations. The help given by Dr Simon Mitton, Science Editorial Director at the Cambridge University Press has been much appreciated.

Finally, to my wife, Beryl, my sincere thanks for her devoted interest and ever-encouraging support at all stages of the work.

1

Introduction to the project – mapping the weather of the past

It is well known that atmospheric circulation is one of the main factors which determine climate. Consequently, changes of climate are indicative of concomitant fluctuations in the flow of the atmosphere. Given adequate sources of data, mapping methods provide the most effective way to make a study of atmospheric circulation.

In order to obtain a better understanding of present atmospheric behaviour, and to become more aware of the kind of changes which could occur in the future, we need to improve our knowledge of past weather and climate on all time scales, both regionally and globally. A detailed study of weather on a daily basis reveals variations in atmospheric behaviour which are not detectable from investigations on longer time scales.

As far as daily records are concerned, it is possible to examine instrumental meteorological observations made in Europe back to the seventeenth century. Documentary sources allow the search for daily wind and weather data to be extended, albeit with decreasing continuity and reliability, back to the Middle Ages. Up to now, however, apart from a specialised case study of weather conditions during the voyage of the Spanish Armada in 1588 (Douglas *et al.*, 1978 and 1979), the analysis of daily circulation patterns by means of synoptic charts* has been made only from the latter part of the nineteenth century, that is, from the time when official observations began to be made and plotted on weather charts by the newly-established national meteorological services. This is too short a period to determine whether or not fluctuations which have occurred in atmospheric circulation over the past century

* A synoptic chart is a bird's-eye view of weather observations over a large area at a specified time.

or so are merely random oscillations in a general trend or are themselves part of a more or less regular cyclic pattern.

Following signs in the 1950s that circulation patterns were beginning to develop differently from those experienced during earlier decades of this century, the idea of extending the synoptic record by constructing daily weather maps, using the mostly untapped sources of daily data from the historical-instrumental period of meteorology, began to take shape. There was no doubt that if realised it would demonstrate that the synoptic reconstruction and analysis of daily circulation patterns could be extended back in time beyond the then existing 100-year record. How did this concept come about?

In a number of lectures and articles presented during the early 1960s, Hubert Lamb actively publicised his belief that mapping methods could make an important contribution to the advancement of the study of climatic change. Over the preceding decade he had been working with his colleague A. I. Johnson at the UK Meteorological Office, reconstructing atmospheric circulation patterns back to 1750 by drawing mean pressure maps for the months of January and July (Lamb and Johnson, 1966). Through handling the large volume of historical weather data brought to light with this research, Lamb realised that many of the sources contained the necessary amount of detail to allow synoptic surveys to be made of daily weather changes.

In his life's work compiling a history of world climate, Lamb had discerned certain periods in the record having such distinctive climatic variations as to warrant further and more detailed investigations using, if possible, daily weather mapping methods. When such a period coincided with the great upsurge of instrumental meteorological observations which occurred during the 1780s, we have both the *raison d'être* – interesting weather and climate, and the necessary ingredients – regionally extensive daily quantitative data, for constructing a series of daily synoptic weather maps.

This then is the philosophy which prevailed in the Lamb school of climatology during the late 1960s. It resulted in a programme of research being undertaken by the Department of Geography under Professor W. G. V. Balchin at the University College of Swansea to construct daily weather maps for the 1780s over the eastern North Atlantic–European sector. The study was initially sponsored by the Meteorological Office for the Meteorological Research Committee, with Professor John Oliver at Swansea liaising with Lamb at the Meteorological Office, Bracknell, the present author becoming involved with the project in 1969. Two years later, following the establishment of the Climatic Research Unit in the School of Environmental Sciences at the University of East Anglia, by Hubert Lamb, now appointed university professor, the research was transferred from Swansea to Norwich.

Placed at a critical phase during the closing stages of the Little Ice Age, the 1780s contain a number of outstanding temperature and rainfall extremes, both positive and negative, which must represent some very pronounced regional anomalies in the general circulation. Then, as now, an apparent increase in the variability of weather and climate from season to season and year to year was causing concern. It appears that as a result of increased volcanic activity from about 1780, repeatedly-replenished dust veils in the upper atmosphere had an important effect in prolonging the Little Ice Age into the first half of the nineteenth century, when purely meteorological trends would otherwise have suggested an earlier reversal to warmer conditions (Lamb, 1970). Similarly today, it seems that weather and climate have again entered a less-stable phase, owing to the complicated interaction of several factors. Basically, it appears that a man-made warming, due to the growth of industrially-produced heat and the increasing concentration of carbon dioxide in the atmosphere, is modulating, to an arguably greater or lesser degree, a cooling trend attributable to natural causes.

The daily weather maps that have been produced in this project are the earliest such charts to be constructed based on quantitative instrumental data from the history of meteorology. Well-established principles of synoptic weather forecasting with regard to the development and movement of pressure systems have been employed in the analysis process to make these historical weather maps directly comparable with current synoptic charts. This research is clearly demonstrating that the daily weather situations which prevailed over Europe two centuries ago can be reconstructed, analysed and classified according to present methods of synoptic meteorology and climatology. The daily continuity which has been maintained in the analysis throughout the series so far has generated sufficient confidence for us to be reasonably sure that the large-scale synoptic features which determined the weather from day to day over Europe during the 1780s can be positively identified and located.

The purpose of this book is to give an account of this research in synoptic climatology. Major results so far achieved are also reviewed, and the maps themselves are presented in a reduced form from the original 26×22 inch working charts. Thus we can see that it is possible to extend the study of daily weather types and circulation patterns back from the mid nineteenth century into the historical-instrumental period of meteorology.

2

The development of observing during the historical-instrumental period of meteorology (1600–1850)

Those who think that meteorology is a comparatively new science are often surprised by the sophistication of the meteorological observations made in the 1780s. But the art of weather observing has a long history.

All down the ages weather has been watched by many people, especially those such as mariners and farmers whose lives and livelihoods depend upon the behaviour of the atmosphere. A large body of empirical knowledge, known as weather lore, gradually evolved, which attempted to relate approaching weather to such items as the appearance of the sky, the behaviour of flora and fauna, and many other natural phenomena. In the early 1600s however, the invention of the barometer and thermometer signalled the transition from purely visual to instrumental observing, through which the study of weather was to be transformed into a more exact and quantifiable science.

Seventeenth-century philosophers showed great interest in these new meteorological instruments, for they appeared to provide the means to investigate causes of weather changes using the scientific method based on systematic observations advocated by Francis Bacon in the early 1600s. It was soon realised that the value of instrumental observations would be greatly enhanced if readings at several different places could be made simultaneously. The earliest documented experiment of this kind was carried out by observers in Paris, Clermont-Ferrand and Stockholm in about 1650.

The first attempt to establish a more permanent network of meteorological stations was made in Italy in 1653, under the patronage of the Grand Duke of Tuscany, Ferdinand II, member of the Medici family and founder of the

Accademia del Cimento (Academy of Experiments) in Florence. Instruments were dispatched to about a dozen stations mostly situated in northern Italy and a uniform procedure for making the observations was devised. This included the recording of pressure, temperature, humidity, wind direction, and state of the sky. The reports were entered on specially prepared forms and periodically sent to the Academy for perusal and analysis. Although the network ceased to function after the Academy was disbanded in 1667, it did set the the pattern for many later attempts.

In fact, the idea of making standardised meteorological observations in concert was taken up in the same year, 1667, by the newly-founded Royal Society of London when Robert Hooke, its first Curator, proposed 'A Method for Making a History of the Weather' (see Figure 2.1.). Hooke himself designed a number of instruments for this purpose including barometers for use on land and at sea, a thermometer graduated with zero corresponding to the freezing point of water, and a pressure-plate anemometer for measuring wind force (see Figure 2.2). The aim to establish a network of meteorological stations was further pursued by the Royal Society in 1723 when its Secretary, James Jurin, issued an invitation to those willing and able to participate in a scheme to make weather observations. Jurin's interest in this undertaking was both meteorological and medical; as a medical student at the University of Leyden he had been influenced by the Dutch physician Hermann Boerhaave, who speculated on the effect weather and climate might have on public health.

The observations organised by the Royal Society in 1723 were recorded in 'Weather Journals' having pages divided into columns for entering the day and hour of the report, the height of the barometric reading in inches and tenths, the temperature in degrees and tenths, the direction and strength of the wind, the state of the sky, a brief description of the weather, and finally the amount of rain or melted snow collected since the previous observation, measured in inches and tenths. The wind strength was estimated on a five-point scale in which 0 denoted a flat calm, 1 a

Figure 2.1. Example from the Royal Society scheme for making weather observations as suggested by Robert Hooke (from Shaw, 1933).

ROYAL SOCIETY. 179

A

SCHEME

At one View repreſenting to the Eye the Obſervations of the Weather for a Month.

Dayes of the Month and place of the Sun. Remarkable houſe.		Age and ſign of the Moon at Noon.	The Quarters of the Wind and its ſtrength.	The Degrees of Heat and Cold.	The Degrees of Drineſs and Moyſture.	The Degrees of Preſſure.	The Faces or viſible appearances of the Sky.	The Notableſt Effects.	General Deductions to be made after the ſide is fitted with Obſervations. As,
14 II 12.46	4 8 12 4 8 12	27 ♉ 9.46. Perigeu.	W 2. 3. 3 1/3 W.S.W. 1	9 3/4 12 1/2 16 10 1/8 7 1/2	2 5 2 8 2 9	29 1/10 29 1/8 29 3/8	Clear blew but yellowiſh in the N. E. Clowded toward the S. Checker'd blew.	A great dew. Thunder, far to the South, A very great Tide.	From the laſt Q. of the *Moon* to the Change the Weather was very temperate, but cold for the ſeaſon; the Wind pretty conſtant between N. & W. A little before the laſt great Wind, and till the Wind roſe at its higheſt, the Quick-ſilver continu'd deſcending til it came very low; after wch it began to reaſcend, &c.
15 II 13.40	8 4 6 10	28 ♉ 24.51.	N. W. 3 4 N. 2 1	9 8 1/2 7	28 1/2 2 9 2 10	29 1/16 29	A clear Sky all day, but a little Checker'd at 4. P. M. at Sun-ſet red and hazy.	Not by much ſo big a Tide as yeſterday. Thunder in the North.	
16 II 14 37	10	N. Moon. at 7. 25' A. M. II 10. 8.	S. 1	10	1 10	28 1/2	Overcaſt and very lowring.	No dew upon the ground, but very much upon Marble-ſtones, &c.	
		&c.	&c.	&c.	&c.	&c.	&c.		

Z 2 DI-

Figure 2.2. Meteorological instruments designed by Robert Hooke for the Royal Society scheme of weather observations (from Shaw, 1933).

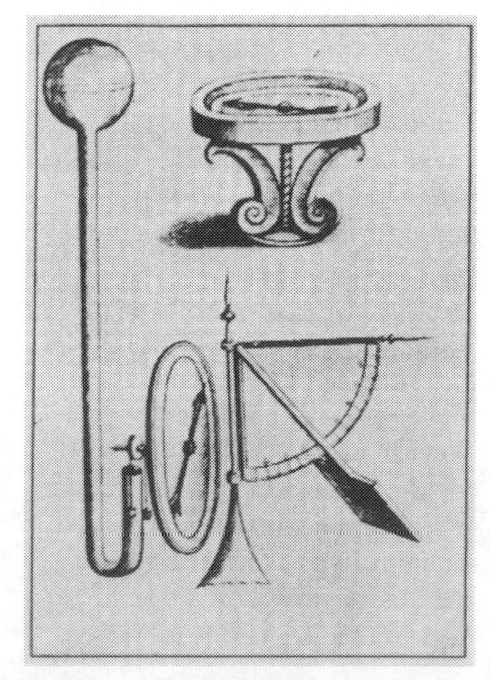

light air, 2 a moderate wind, 3 a strong wind, and 4 a most violent wind. Observers were requested to send copies of their journals every year to the Royal Society in London for publication in the *Philosophical Transactions*, and for a few years reports were received from correspondents in Europe, America and Asia.

Other scientific bodies and individuals attempted to organise networks of weather observing stations in the eighteenth century, including the Great Northern Expedition to Siberia under Bering in the 1730s, and Goethe in the German state of Sachsen-Weimar during the 1770s. However, these endeavours do not provide the geographical coverage nor continuity of data required for any sustained synoptic study, and it is not until the 1780s that sufficient material becomes available for this purpose. During this decade concerted efforts to make and collect standardised meteorological observations were begun on a grand scale in both France and Germany, and it is these which provide the core of data for the present investigation.

Following the lead of meteorologically-inclined physicians, such as Boerhaave and Jurin, medical authorities in France during the late eighteenth century decided to make a systematic study of the possible relations between weather conditions and public health. Consequently, in 1778 the *Société Royale de Médecine* was established, under the patronage of Louis XVI, to maintain a regular and ordered correspondence on meteorological and medical matters with doctors throughout the kingdom (Kington, 1970). Vicq d'Azyr was appointed Secretary-General and, together with the French meteorologist-cleric, Le Père Louis Cotte, became actively involved with establishing and maintaining an extensive network of weather stations for the Society with the country's physicians acting as observers. Four years earlier in his *Traité de Météorologie* Cotte had given information about how to make daily instrumental observations, based on his own experience in keeping a weather journal at Montmorency near Paris. Similar instructions on instrumental exposure, observational procedure, and methods of recording and summarising the daily reports were now issued by the Society so as to ensure that standardised, comparable observations would be made by the team of doctors. The correspondents were requested to make observations of pressure, temperature, wind, humidity, rainfall, evaporation, state of the sky, and significant weather three times a day at sunrise ('matin'), about two to three hours after midday ('midi') and at 9 or 10 o'clock in the evening ('soir'). They were issued with specially printed forms for recording the daily reports in monthly sets (see Figure 2.3) which were regularly sent to the Paris headquarters of the Society for collation.

Whilst many eighteenth-century observers had previously taken readings of temperature simply from indoor thermometers hung in fireless rooms, a serious attempt was now made to obtain a more representative and comparable value by exposing thermometers in sheltered north-facing sites outside. The instruments were graduated in degrees Réaumur, a scale on which the freezing point and boiling point of water are 0 ° and 80 ° respectively. Observations of atmospheric pressure were made with mercury barometers graduated in Paris inches and lines. Before the adoption of the metric system during the Napoleonic era, Paris inches and lines were commonly used in Europe for measuring atmospheric pressure. The Paris inch was divided into 12 lines, each line being further subdivided either into tenths or twelfths. One Paris inch is equivalent to a pressure interval of 36.1 millibars, i.e. one Paris line is approximately equal to 3.01 millibars. Where possible, vernier-type devices were fitted to the barometers to allow readings to be made to an accuracy of tenths or twelfths of a line, that is, thirds or quarters of a millibar.

By the mid 1780s the observation network comprised over 70 stations and had been extended beyond the borders of France to include correspondents in other countries of Europe as well as in America and Asia (see Table 2.1). This was the kind of meteorological organisation envisaged by the French scientist Antoine Lavoisier in 1781 when he pointed out that in order to prepare a forecast the meteorologist would need simultaneous daily observations of various weather elements:

> The forecast of the changes due to the weather is an art that has its principles and

Figure 2.3. Register of daily observations for October 1781 kept at Dijon, Burgundy by Dr Maret for the *Société Royale de Médecine.* Thrice-daily readings are given of barometer (Paris inches, lines and twelfths), thermometer (degrees Réaumur), wind direction, state of the sky, and significant weather. Monthly summaries relating to the atmosphere, botany and diseases are also given (from Kington, 1980).

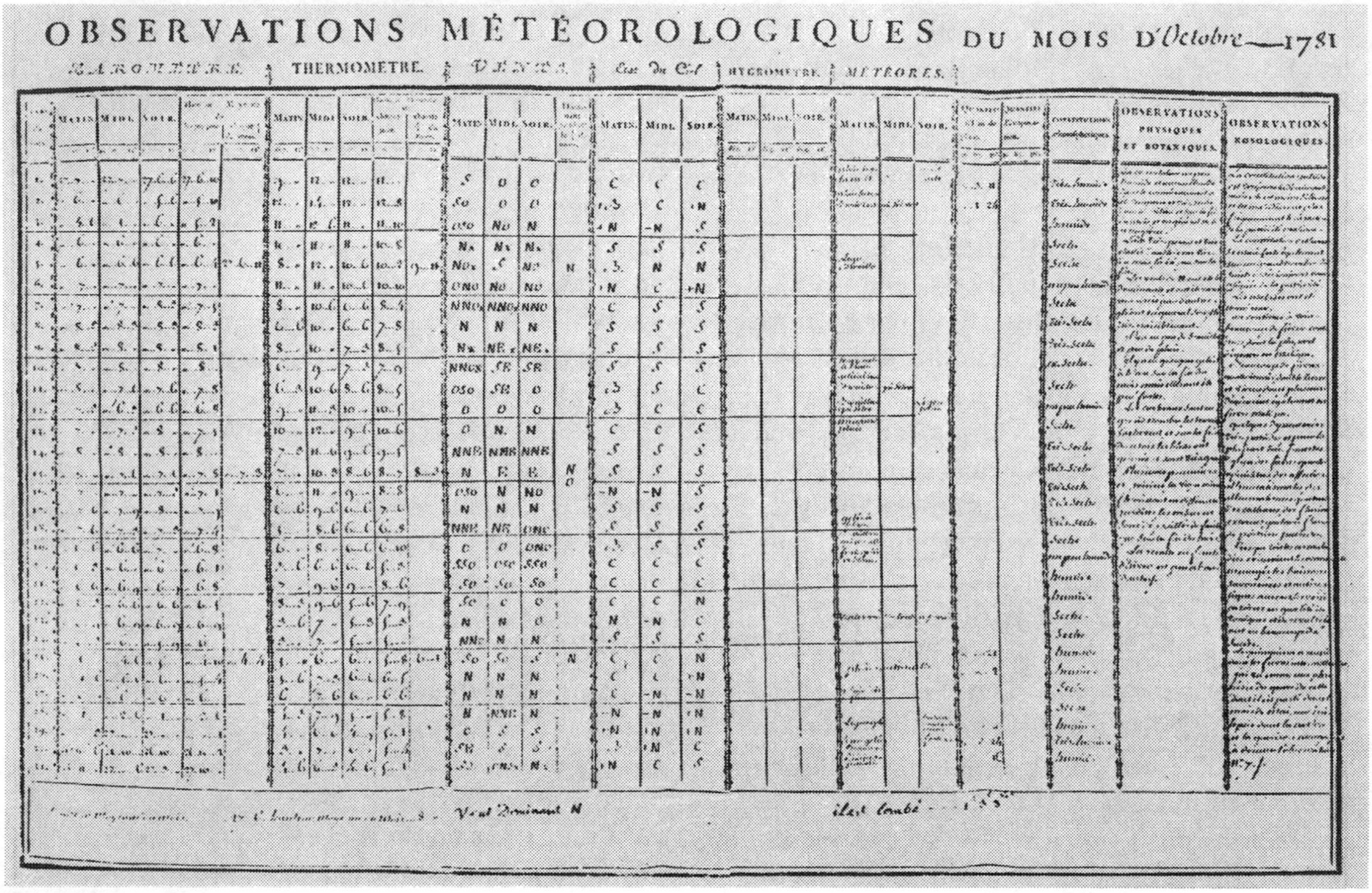

OBSERVATIONS MÉTÉOROLOGIQUES DU MOIS D'Octobre—1781

BAROMETRE | THERMOMETRE | VENTS | État du Ciel | HYGROMETRE | METEORES

OBSERVATIONS PHYSIQUES ET BOTANIQUES | OBSERVATIONS NOSOLOGIQUES

Vent Dominant N

[illegible]

Table 2.1. *Catalogue of meteorological stations making weather observations in the 1780s.*

Key
SRM: *Société Royale de Médecine*
SMP: *Societas Meteorologica Palatina*
p: pressure t: temperature u: humidity w: wind sky: state of the sky wr: weather r: rain
e: evaporation

Station	Period	Observer	Authority	Observation Times	Elements Observed
Albury, Surrey	1782–96	Godschell Man	Private	1×daily	p, t, w, wr
Ancenis, Brittany	1786–92	—	SRM	3×daily	p, t, w, sky, wr
Argentat, Limousin	1783–5	Lertourgie	SRM	3×daily	p, t, sky, wr
Arles, Provence	1782–8	Dr Bret	SRM	3×daily	p, t, w, sky, wr, r
Arras, Artois	1783–93	Dr Buissart	SRM	3×daily	p, t, u, w, sky, wr
Aylsham/Buxton, Norfolk	1786–91	A. E. Powell	Private	2×daily, probably at 0900 and 2100 with occasional readings at noon	p, t, w, wr, r
Barcelona	1780–1827	F. Salva y Campillo	Private	0600, 1400, 2300	p, t, w, sky, wr

Table 2.1. (*cont.*)

Station	Period	Observer	Authority	Observation Times	Elements Observed
Basel	1755–94	—	—	0700, 1300, 2100	p, t (mean daily values), wr
Beaune, Burgundy	1784–5	Lecousturier de l'Oratoire	SRM	3×daily	p, t, w, sky, wr
Belmont Castle, Perthshire	1782–99	J. S. Mackenzie	Private (Marquis of Bute)	2×daily	p, t, w, wr, r
Berlin	1781–8	N. von Beguelin, Professor, Academy of Sciences	SMP	0700/0800, 1430, 2200	p, t, u, w, sky, wr
Bern	1760–	—	—	daily	p, t (mean daily values), wr
Besançon, Franche-Comte	1783–7	Meillardet	SRM	3×daily	p, t, w, sky, wr
Billom, Auvergne	1779–90	Avinet	SRM	0600–0800, occasionally 1500	p, t, w, sky, wr
Bologna	1782–92	Matteuci, Professor of Astronomy and Physics	SMP	0700, 1400, 2100	p, t, u, w, sky, wr
Brest, Brittany	1783–4	Aublct	SRM	3×daily	p, t, w, sky, wr
Bridestow, Devon	1782–4	T. Heberden	Private	0800, 1400	p, t, wr
Brussels	1782–92	Mann and Chevalier, Abbots, Academy of Sciences	SMP	0700, 1400, 2100	p, t, w
Buda	1781–92	F. Weiss (Astronomer Royal) and F. Bruna	SMP	0700, 1400, 2100	p, t, u, w, sky, wr, r
Cadiz	1786–9	—	Private	daily	p, t, w, sky, wr
Cambuslang, Lanarkshire	1785–1809	J. Meek, D.D.	Private	0800, 1400/1500, 2200	p, t, w, sky, wr, r
Castel Sarrazyn, Languedoc	1783–5	Ressayre	SRM	3×daily	p, t, w, sky, wr
Châlons-sur-Marne, Champagne	1785–90	Dr Moignon	SRM	3×daily	p, t, w, sky, wr
Chinon, Touraine	1780–91	Linacier	SRM	3×daily	p, t, w, sky, wr
Clermont Ferrand	1784–90	—	SRM	3×daily	p, t, w, sky, wr
Copenhagen	1781–90	T. Bugge, Astronomer Royal; Professor of Astronomy and Mathematics, University of Copenhagen	SMP	0700, 1200, 2100	p, t, u, w, sky, wr, r
Dax, Gascogne	1782–5	Dufan	SRM	3×daily	p, t, w, sky, wr
Delft/The Hague	1782–6	S. P. van Swinden	SMP	0700, 1400, 2100	p, t, u, w, wr, r, e
Dijon, Burgundy	1761–85	Dr Maret	Private, SRM and SMP	3×daily	p, t, w, sky, wr, r, e
Downing Hall, Holywell, Flintshire	1784–90	T. Pennant	Private	daily, mostly	p, t, w, wr

Table 2.1. (*cont.*)

Station	Period	Observer	Authority	Observation Times	Elements Observed
Dublin	1781–1811	M. Faviere, Comptroller of the Works, Dublin Castle	Private	daily	wr
Dublin	1783–4	—	Private	2×daily, mostly	p, t, w, sky
Düsseldorf	1782–4	Liessen and Phennings	SMP	0700, 1400, 2100	p, t, w, sky, wr
Epoisse, Burgundy	1784–5	Forestier	SRM	3×daily	p, w, sky, wr
Erfurt	1781–8	Planer	SMP	0700, 1400, 2200	p, t, w, sky, wr, r, e
Eyjafirth, Iceland	1747–1846	J. Jónsson (father and son)	Private	Daily	w, sky, wr
Faeroes	1781–2	—	Private	3×daily	p, t, w, sky, wr
Geneva	1782–9	Senebier	SMP	0700, 1300, 2100	p, t, u, r
Gordon Castle, Morayshire	1781–1827	J. Hoy	Private	0800	p, t, w, wr
Göttingen	1783–7	Gatterer	SMP	0700, 1400, 2100	p, t, u, w, sky, wr, r, e
La Grande Combe des Bois, Franche-Comte	1783–8	Mougin Lude	SRM	0800	p, t, u, w, sky, wr
Grenoble, Dauphiné	1783–4	Dr Chanoine	SRM	3×daily	p, t, u, w, sky, wr
Haguenau, Alsace	1781–91	Dr Keller	SRM	3×daily	p, t, w, sky, wr
Halifax, Yorkshire	1782–6	C. Ashworth	Private	daily	wr
Hamburg	1767–1808	Brodhagen	Private	3×daily	p, t, w, wr
Harleston, Norfolk	1786–1832	T. Passant	Private	daily	w, wr
Herbipolis (Würzburg), Franconia	1781–8	Egel	SMP	0700, 1400, 2100	p, t, u, w, sky, wr, r
Johnstone, Renfrewshire	1768–1805	J. Houston	Private	daily, mostly	wr
Kemnay/Disblair, Aberdeenshire	1758–95	J. Burnett	Private	daily, mostly	w, wr
Kincardineshire	1782	G. Watson	Private	daily	w, wr
Laigle, Normandy	1784–5	—	SRM	3×daily	p, t, w, sky, wr
Lambhús, Iceland	1779–89	R. Lievog, Astronomer	Royal Danish Government	0600, 1200, 1800, 2000/2100	p, t, w, sky, wr
Leningrad	1782–92	Euler, Secretary, Academy of Sciences	SMP	0600, 1200/1400, 1800/2200	p, t, w, sky, wr
Lille, Flanders	1783–6	Dr Saladin	SRM	3×daily	p, t, u, w, sky, wr r, e
Liverpool (Old Dock Gates)	1768–93	W. Hutchinson	Private	0800, 1200	p, t, w, sky, wr
London	1723–1805	T. Hoy *et al.*	'London Weather Diary' (Manley)	daily	p, t, w, wr
Lons-le-Saunier, Franche-Comte	1784–90	—	SRM	3×daily	p, t
Luçon, Poitou	1785, 1788	Merland de Chaille	SRM	3×daily	p, t, w, sky, wr
Lyndon Hall, Rutland	1748–63; 1777–89	T. Barker, Country Squire	Private	0700, 1400	p, t, w, wr, cloud motion, r
Madrid	1784–9	—	Private	0800	p, t, w, sky, wr

Table 2.1. (*cont.*)

Station	Period	Observer	Authority	Observation Times	Elements Observed
Mafra, Portugal	1783–6	D. Joaquim da Assumpção Velho, Professor of Physics and Mathematics	Real Collegio de Mafra	3×daily	p, t, w, sky, wr, r
Mannheim	1781–92	J. Hemmer	SMP	0700, 1400, 2100	p, t, u, w, sky, wr, r
Mar, Aberdeenshire	1783–92	Earl of Fife	Private	daily (August, September and October)	wr
Marseilles	1782–92	S. Jacques de Silva Belle	SMP	0800, 1400, 2200	p, t, u, w, sky, wr
Metz, Lorraine-Moselle	1783–4; 1790–1	Gentil (1790–1)	SRM	3×daily	p, t, u, w, sky, wr
Middelburg, Zeeland	1782–8	J. A. van de Perre	SMP	0700, 1400, 2100	p, t, u, w, sky, wr r
Milan	1763–1838	F. Reggio	Private	2×daily	p, t, w, sky, wr,
Minehead, Somerset	1783–4	J. Atkins	Private	0600/0800, 1200, 2100	p, t, w, wr, r
Mirebau, Anjou	1780–7	Ayrault	SRM	3×daily	p, t, w, sky, wr
Modbury, Devon	1788–1868	J. Andrews (father and son)	Private	daily; 0800, 2000 from 11/1790; 0800, 1200, 2000 from 1/1791	wr
Mongewell, Oxford	1771–1823	Rev. S. Barrington Bishop of Durham	Private	0800, 1200, 2200	p, t, w, wr
Montdauphin, Dauphiné	1783–8	D'Arbalestier and Charmeil (Surgeon Major)	SRM	3×daily	p, t, w, sky, wr
Montdidier, Picardy	1783–1869	Drs V. and C. Chandon	SRM and private	3×daily	p, t, u, w, sky, wr
Montlouis, Roussillon	1780–5	Dr Barrère	SRM	3×daily	p, t, w, sky, wr
Montluçon, Bourbonnais	1784–5	Rochette	SRM	3×daily	p, t, w, sky, wr
Moscow	1783–9	Engel and Stritter	SMP	0600, 1400, 2200	p, t, w, sky, wr
Mulhausen, Alsace	1778–86	D. Meyer	SRM	3×daily	p, t, u, w
Munich	1781–92	Huebpauer, Reader in Theology, Augustine order	SMP	0700, 1400, 2100	p, t, w, sky, wr
Nantes, Brittany	1779–83	Richard du Plessir	SRM	3×daily	p, t, w, sky, wr
Norwich	1778–1815	W. Youell, Manager, New Mills	Private	2×daily	w, wr
Padua	1781–92	Toaldo	SMP	0700, 1400, 2100	p, t, w, sky, wr, r
Perpignan, Roussillon	1783–5	—	SRM	3×daily	p, t, w, sky, wr
Poitiers, Poitou	1775–1819	Dr de la Mazière	SRM and private	3×daily	p, t, w, sky, wr

Table 2.1. (*cont.*)

Station	Period	Observer	Authority	Observation Times	Elements Observed
Pontarlier, Franche-Comte	1777–5, 1791–3	Dr Tavernier and le Père F. Tavernier	SRM	3×daily	p, t, w, sky, wr
Prague	1781–91	A. Strnadt, Astronomer Royal	SMP	0700, 1500, 2100	p, t, u, w, sky, wr, r
Regensburg	1781–91	P. Heinrich, Cleric and Professor of Experimental Physics	SMP and SRM	0700, 1400, 2100	p, t, u, w, sky, wr, r
Ribe, Denmark	1786–1816	L. Heinssen	Private	daily, mostly	wr
Rieux Evêché, Languedoc	1783–90	Abbot Darbas	SRM	3×daily	p, t, w, sky, wr
La Rochelle, Aunis	1777–93	Seignette, Secretary, Academy of Sciences	SMP and SRM	0700, 1400, 2100	p, t, w, sky, wr, r, e
Rome	1782–92	Abbot G. Calandrelli, Professor of Mathematics, College of Rome	SMP	0700, 1400, 2100	p, t, u, w, sky, wr, r, e
Rouen, Normandy	1777–80, 1783–90	Lépecq de la Clôture	SRM	3×daily	p, t, w, sky, wr
St Brieuc, Brittany	1783–5	Dr Bagot	SRM	3×daily	p, t, w, sky, wr
St Dié, Lorraine	1783–9	Dr Félix Poma and G.-F. Renaud	SRM	3×daily	p, t, w, sky, wr
St Gotthard	1781–92	P. Laurentio Mediolanensi and Onuphrius, Franciscian Order	SMP	0700, 1400, 2100	p, t, u, w, sky, wr
St Jean d'Angély, Aunis	1786–9	—	SRM	0600, 1400, 2200	p, t, w, sky, wr
St Lô, Normandy	1786–	—	SRM	3×daily	p, t, w, sky, wr
St Malo, Brittany	1782–8	—	SRM	3×daily	p, t, u, w, sky, wr
St Paul-Trois-Châteaux, Dauphiné	1781–91	Dr Caudeiron	SRM	3×daily	p, t, w, sky, wr
Selborne, Hampshire	1768–93	Rev. G. White	Private	0800, 2000	p, t, w, wr
Skálholt, Iceland	1777–81	H. Finnsson, Bishop	Private	1000, 2200	p, t, w, sky, wr
Spitzbergen, Norway	1783–6	Wilse	SMP	0700, 1400, 2100	p, t, w, sky, wr
Stockholm	1783–7	P. Wargentin and H. Nicander, Secretaries, Academy of Sciences	SMP	0600/0700, 1300/1400, 2200	p, t, u, w, sky, wr
Stroud, Gloucestershire	1771–1813	Dr T. Hughes	Private	0700, 1400, 1800 and 'night'	p, t, w, sky, wr

Table 2.1. (*cont.*)

Station	Period	Observer	Authority	Observation Times	Elements Observed
Thorseng, Denmark	1779–88	—	Private	0700, 1200, 2200	t, w, sky, wr
Tournus, Burgundy	1784–93	Dr Dunard	SRM	3×daily	p, t, w
Trondheim	1762–1802	J. Berlin and D. Fester	Private	1200, 2300/2400	p, t, w, sky, wr
Troyes, Champagne	1778–86	Le Père le Bouthilier and Le Père Rondeau	SRM	3×daily	p, t, u, w
Truro/Redruth, Cornwall	1785–8	S. James	Private	0800, 1400	p, t, w, wr
Vabres, Guyenne	1784–5	Mabrieu	SRM	3×daily	p, t, w
Vannes, Brittany	1784–7	Dr Aubry	SRM	3×daily	p, t, u, w
Vienna	1775–1855	—	Carl von Littrow k.k. Sternwarte	0800, 1500, 2200	p, t, (w, sky, wr from 1790s)
Vienne, Dauphiné	1778–83	Revolat	SRM	3×daily	p, t, w, sky, wr
Warsaw	1779–1828	J. Bończa-Bystrzycki (1779–99)	Royal Astronomical Observatory	3×daily	p, t, w, wr
Zagan, Lower Silesia	1781–92	Presus, Augustine Order	SMP	0700/0800, 1300/1400, 2100	p, t, u, w, sky, wr, r, e
Zwanenburg, Holland	1735–	—	—	0700/0800, 1400, 1900	p, t, w, sky, wr, r

its rules, that demand a great deal of experience and the attention of a very skilled physicist. The required data for this art are: the regular and daily observations of the variations of the height of the mercury in the barometer, the force and direction of the winds at different levels, the hygrometric conditions of the air . . . . With all these data it is virtually always possible to forecast a day or two in advance how the weather is most likely going to be: it is even thought that it would be possible to publish a forecast bulletin every morning, which would be of great advantage to society.

Unfortunately, as we shall see in more detail later, the means to transmit the observations to a central point where they could be processed rapidly enough to keep pace with the development and movement of transitory and migratory weather systems were lacking at that time. Although meteorologists had to wait several more decades for advances to be made in both communications and in synoptic weather studies before Lavoisier's dream could be realised, it is clear that the scientists associated with the *Société Royale de Médecine* had recognised the importance of making simultaneous daily observations over a large area for the purpose of carrying out comparative surveys on the weather. Sadly, this enlightened organisation was suppressed by a Revolutionary decree of the new French state in 1793, and the following year Lavoisier was guillotined. However, the original manuscripts containing daily observations made in the 1780s were preserved in the archives of the Académie de Médecine in Paris, where after nearly two centuries of oblivion they were eventually brought to light in 1965 by the French historian Emmanuel Le Roy Ladurie and his research team at the Sorbonne (Desaive *et al.*, 1972).

This was not the only scheme afoot in the 1780s for the advancement of meteorology. Mannheim, the capital of the eighteenth-century German state, the Rhineland Palatinate, had developed into an influential centre of the arts and sciences during the Enlightenment. In 1780, Karl Theodor, Prince-Elector of the Palatinate, founded the *Societas Meteorologica Palatina* and appointed his court chaplain, Johann Hemmer, as its director (Kington, 1974). This was the first purely meteorological society to be established with the primary object of predicting the weather by analysing data collected from a network of stations making systematic daily observations.

Each of the Society's observers was supplied *gratis* with a set of uniformly calibrated instruments together with detailed instructions, written in Latin by Hemmer, on observational procedure. The mercury barometer was graduated in Paris inches and lines with a vernier scale (see Figure 2.4); it was to be firmly suspended on a wall of a room in the observatory, and be unaffected by undue changes of temperature or direct sunlight. The thermometers were of the mercury-in-glass pattern, and graduated in degrees Réaumur (see Figure 2.4). One thermometer was to be mounted indoors adjacent to the barometer so that pressure readings could be corrected for the temperature of the mercury. A second thermometer was to be mounted outside the observatory in a sheltered north-facing exposure, not affected by the sun's rays, in order to record air-shade temperature. The observer at Prague, Antony Strnadt, kept his thermometer in a perforated wooden box opening towards the north, a prototype of the now standard equipment for housing meteorological thermometers, the Stevenson screen. Like the *Société Royale de Médecine*, correspondents were requested to make observations three times a day at 0700 h, 1400 h and 2100 h of pressure, temperature, wind, humidity, rainfall, evaporation, state of the sky, clouds, and significant weather. The meteorological registers were regularly dispatched to Mannheim for collation and publication in the *Ephemerides* of the Society, which were provided free to all participating observers. From the entries in these yearly publications (see Figure 2.5) it can be seen that the Society was using a system of weather reporting symbols which owed something of its origin to an earlier scheme devised by Pieter van Musschenbroek; traces of it still survive in the present international synoptic weather code.

Figure 2.4. Mercury barometer and mercury thermometer of the *Societas Meteorologica Palatina* (from Kington, 1974).

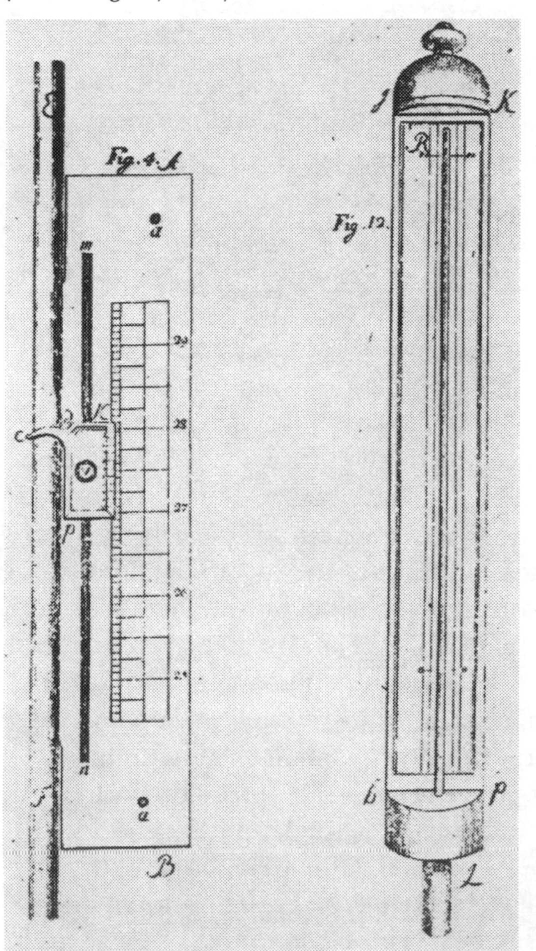

The wind force scale of the Society (see Table 2.2.) was a development of the one that had been adopted by Jurin in his scheme for the Royal Society in 1723. This type of scale was used by many weather observers in the eighteenth century for estimating wind strength. It is interesting to note that when the numbers are doubled, the corresponding specifications are identical to those used today for describing the effect of wind force over land areas in the Beaufort scale. No doubt Admiral Francis Beaufort was aware of the terminology when he introduced his own scale in 1805.

From a nucleus of about a dozen stations, mostly located in central Europe, the Palatinate network rapidly spread, so that by the mid 1780s it included over 50 observatories extending from Siberia across Europe to Greenland and eastern North America. The principal stations and correspondents of the Society are given in Table 2.1(p. 6). The observers, mostly mathematicians, physicists, astronomers and clerics, were associated with the scientific academies and learned societies established in many European cities during the Enlightenment.

The *Societas Meteorologica Palatina* made an important contribution to the development of meteorology by the effective way it integrated, refined and authoritatively published many of the improvements made in meteorological observing and instrumental design during the seventeenth and eighteenth centuries. It also introduced several of its own ideas in these fields, such as the

Figure 2.5. Extract from the *Ephemerides* of the *Societas Meteorologica Palatina*, showing daily observations made at Mannheim by Johann Hemmer 1–15 January 1785. The columns contain thrice-daily readings (0700 h, 1400 h and 2100 h) of barometer (Paris inches, lines and tenths), interior and exterior thermometers (degrees Réaumur), hygrometer, magnetic declination, wind velocity, state of the sky , and significant weather. Rainfall, height of the Rhine and phases of the moon were also regularly recorded.

OBSERVATIONES MANHEIMENSES

Autore HEMMERO.

Horae observationis ordinariae 7 mat. 2 pom. 9 vesp.

Januarius.

Dies.	Barom.	Th. int.	Th ext. I	Th ext. II	Hygr.	Declin.	Ventus.	Pluvia.	Evap.	Rhen.	Luna.	Coeli fac.	Meteora.
	lg. lin. dec.	gr. dec.	gr dec.	gr. dec.	gr dec.	gr. min.	direct. vires.	p. lin. lignee	lin. dec	ped. dig.			
1	27, 2, 3 1, 8 2, 1	1, 3 1, 6 1, 8	—2, 6 —1, 6 —2, 0		20, 0 22, 5 18, 8	19, 33 45 36	N O 1.1/2 N N O 1 N N O 1	18		-7, 1	♍ ⊙ per.	== == neb. ==	aër nebulos. ∷ t. aliquoties per diem.
2	27, 1, 7 1, 1 1, 0	1, 6 1, 8 1, 8	—1, 5 —0, 8 0, 0		18, 2 21, 3 17, 7	19,37.1/2 48 45	N O 1 N O 1 N N O 1	16		-7, 1	♎	== == ==	∷ t. h. 6 m. & 8 vesp. regelat.
3	27, 1, 1 1, 7 3, 0	1, 8 2, 3 2, 5	—0, 8 1, 9 2, 2		16, 0 18, 8 17, 3	19, 42 46 45	N N O 1 O S O 1 S O 1.1/2	234		-6, .7	☾ h. 7 m. 35 vesp. ♎	== == ==	regelat.
4	27, 3, 5 5, 0 6, 6	2, 8 3, 3 3, 3	2, 4 3, 4 3, 7		18, 6 19, 3 15, 8	19, 45 46 30	S O 1 S S O 1.1/2 O S O 1			-6, 11	♎	== == =	regelat.
5	27, 6, 5 5, 6 5, 7	3, 5 3, 8 4, 0	3, 8 5, 4 5, 9		20, 0 28, 3 30, 0	19, 38 47 47	S O 2 O N O 1 S O 2			-6, 11	♏	== == ++ pall.	regelat.
6	27, 5, 4 4, 2 4, 8	4, 0 4, 6 4, 8	5, 8 8, 2 6, 3		29, 3 36, 2 23, 2	19, 33 47 33	N N O 2 O S O 2.1/2 S S O 1.1/2			-6, 6	♏		
7	27, 8, 2 9, 3 10, 3	4, 8 4, 8 4, 8	1, 3 3, 0 —0, 3		23, 5 30, 4 26, 0	19, 41 46 39	N N W 1 N N W 1.1/2 N O 1			-5, 6	♐	in Ost a. sp. ⊙+	∵ ad montes noct post2proc. ex N N W
8	27, 11, 6 28, 0, 7 1, 3	4, 4 4, 0 4, 0	—2, 5 0, 6 0, 8		21, 4 25, 3 23, 2	19, 39 47 33	N W 1.1/2 O S O 1 S S W 1.1/2			-4, 0	♐	fasc. ⊙ ==	∵ fortis ∷ cop. vesp. ∵ ad mont.
9	28, 1, 7 2, 8 4, 0	4, 0 4, 3 4, 2	—0, 8 1, 8 —1, 2		16, 0 20, 4 22, 5	19, 36 48 34.1/2	N O 1.1/2 N O 1 O N O 1	151		-3, 6	♑ ☽	== ⊙	∷ noct. praec. ∵ h. 9 mat. & ad mont. tot. die.
10	28, 2, 8 1, 9 1, 1	3, 3 3, 5 4, 7	—3, 5 —0, 6 —2, 3		23, 4 26, 4 23, 7	19, 44 45 45	O N O 2 N N W 1 N N W 1			-4, 0	♑	⊙ ⊙ ⊙	∵ ad mont. tot. die aër neb. a prandio.
11	27, 11, 9 11, 2 10, 8	3, 5 3, 3 3, 4	—5, 3 —3, 0 —5, 2	—7, 2	17, 7 21, 7 17, 2	19, 45 45 45	S O 1. W 1 W N W 1			-4, 6	🌕 h. 1 m. 55 mane. ♒	⊙+ ⊙+ ⊙+	∵ ∵ h. 7.1/2 man. aër neb. a meridie.
12	27, 10, 1 9, 5 9, 5	2, 8 2, 8 2, 8	—7, 5 —4, 2 —6, 8	—8, 8	17, 4 21, 7 17, 4	19, 45 48 41	O S O 1 S S W 1 N N W 1			-5, 0	♒ ☾	nebul. ⊙ nebul.	∵ sp. & ∵ fort. item ∵ vesp.
13	27, 9, 4 9, 1 9, 1	2, 2 2, 3 2, 4	—7, 5 —3, 4 —2, 3	—8, 6	17, 0 21, 2 19, 4	19, 41 44 39	N N W 1 N 1 N N W 1			-5, 7	♓	nebul. nebul. == t.	∵ & ∵
14	27, 8, 8 8, 5 8, 5	2, 5 2, 7 2, 7	—1, 6 1, 6 —1, 2		18, 5 23, 4 15, 9	19, 38 51 43	N N O 1 N N W 1.1/2 N 1			-6, 0	♓	== t. == t. ⊙+	
15	27, 8, 2 8, 6 9, 3	2, 7 2, 7 2, 7	—3, 3 0, 5 0, 2		16, 8 21, 5 20, 4	19, 41 52 46	N O 1 O 1 N N O 1			-6, 1	♈	c. in O == ==	∵ t. h. 8 man. h. 8.1/2 aër neb. tot. die

Ephemer. anni 1785. A

Table 2.2. Societas Meteorologica Palatina *(SMP) specifications and plotting symbols for wind force together with Beaufort scale equivalents.*

Venti vires/Wind Strength						
SMP no.	SMP	Beaufort	Description	Beaufort force	Knots (approx.)	Plotting symbol
0		Smoke rises vertically	Calm	0	0	◎
½		Wind direction shown by smoke drift, but not by wind vanes	Light air	1	2	—o
1	*Arborum duntaxat folia*	Leaves rustle	Light breeze	2	5	[illegible]
1½		Leaves and small twigs in constant motion	Gentle breeze	3	9	[illegible]
2	*Ramos minoris agitat*	Small branches are moved	Moderate breeze	4	13	[illegible]
2½		Small trees in leaf begin to sway	Fresh breeze	5	18	[illegible]
3	*Ramos majoris agitat*	Large branches in motion	Strong breeze	6	24	[illegible]
3½		Whole trees in motion	Near gale	7	30	[illegible]
4	*Ramos avellit*	Breaks twigs and boughs of trees	Gale	8	37	[illegible]

first attempt to make systematic reports on cloud forms foreshadowing the now more well-known classification devised by the English meteorologist Luke Howard in 1803. Occasional articles based on the collected daily meteorological observations were also published in the Society's *Ephemerides*, such as the discussion on the persistent high-level dust haze which occurred in 1783 following the exceptional volcanic activity in Iceland and Japan of that year.

Although Hemmer died in 1790 the activities of the Society continued for a further five years. But, unfortunately, the publication of the twelfth volume of the *Ephemerides*, containing data for 1792, brought the series to a close. The Society was facing increasing financial problems and the final blow which led to its disbandment was the fall of Mannheim to French Revolutionary forces in 1795.

Besides the 80 or so weather reporting stations located in Europe in the networks organised by the two European scientific societies, a large number of private observers were recording daily meteorological observations during the eighteenth century (see Table 2.1 again). In the British Isles these more individualistic efforts were mostly made by physicians, country parsons and landed gentry. Although working in isolation these people sometimes corresponded with one another about their mutual interests in meteorology and natural philosophy, the Royal Society in London providing a centre for the more formal discussion and exchange of ideas. In fact, the efforts of Robert Hooke and James Jurin to establish meteorological stations for the Society in the 1660s and 1720s had undoubtedly not been forgotten, and by the 1780s comparable methods of recording weather elements at more or less standard times every day were being carried out at a number of places in the British Isles. One of the best examples of these individual eighteenth-century meteorological observers was the Rutland squire, Thomas Barker, who, beginning in the 1730s, kept a weather journal for over 60 years at Lyndon Hall (see Figure 2.6). There were also many individual weather observers in other parts of Europe. For example, on the instigation of the Danish government the Scandinavian astronomer

Rasmús Lievog set up a weather station at Lambhús in Iceland, where he made instrumental observations (see Figure 2.7) for a number of years in the 1780s (Kington, 1972).

All these efforts strongly demonstrate that collectively or individually the need to establish a good observational data base in order to gain a better knowledge and understanding of weather and climate was generally recognised during the closing years of the Enlightenment. We are indeed fortunate that due to the painstaking and meticulous endeavours of many learned people of that time we can now make a detailed study of the weather of the 1780s over Europe.

Before closing this general account of historical meteorological observations there remains one further source of weather data to be mentioned. The archives of the Public Record Office in London and the National Maritime Museum in Greenwich house large collections of British naval logbooks. The weather reports contained in these logbooks may not be instrumental but, made with a disciplined experience of watch-keeping dating back to the 1670s, they do provide us with a vast amount of useful and quantifiable information about daily wind and weather conditions over sea areas and in coastal regions of Europe and beyond throughout the eighteenth century (Oliver and Kington, 1970).

Entries were made in logbooks every day at noon for the previous 24 hours, this period often being divided into three parts: first, middle and latter, each comprising two ships' watches. The ship's position at noon was given each day; when at sea, the latitude and longitude in degrees and minutes; if in harbour, by the name of the port or anchorage. The log entry always included details

Figure 2.6. Extract from the meteorological register kept at Lyndon Hall, Rutland by Thomas Barker for June and July, 1783. The two pages illustrated from this manuscript record show daily observations made at about 0700 h and 1400 h of pressure in English inches and hundredths, temperatures in degrees Fahrenheit (interior and exterior thermometers), cloud velocity, wind velocity, rain, state of the sky, and significant weather (from Kington, 1980).

Table 2.3. *Terms used to describe state of the sky, weather and visibility in ships' logbooks during the eighteenth century.*

State of the sky	
Serene	Cloudy
Clear	Overcast
Fine	Thick weather
Fair	
Weather	
Drizzle	Shower
Rain	Rain squalls
Snow	Thunderstorm
Sleet	Lightning
Hail	
Visibility	
Haze	Fog
Thick haze	Thick fog
Mist	

about the weather, the most important items being wind direction and strength. Wind direction was recorded on the 32-point compass, with significant changes over the 24-hour period being shown by a sequence of entries (see Figure 2.8). Estimates of wind strength were probably related to the effect of wind pressure on ships' sails and the surface of the sea. The reports show that eight expressions were being used in a fairly consistent way, namely, 'calm', 'light air', 'light breeze', 'moderate breeze', 'fresh breeze', 'fresh gale', 'strong gale' and 'hard gale', which can be related to the present Beaufort wind force scale. General descriptive remarks about the weather were also given and, as with the wind force terms, it is again evident that a standardised set of terms was being used (see Table 2.3).

Figure 2.7. Extract from the meteorological register kept at Lambhús, Iceland by Rasmús Lievog in May 1782. The two pages illustrated from the manuscript record are of observations made in the morning (0600 h) and at midday (1300 h). The columns contain readings of barometer (Paris inches, lines and quarters or sixths), exterior thermometer (degrees Réamur), wind direction, state of the sky, and significant weather. This last column also contains standard terms used by Danish observers to describe wind strength (from Kington, 1972).

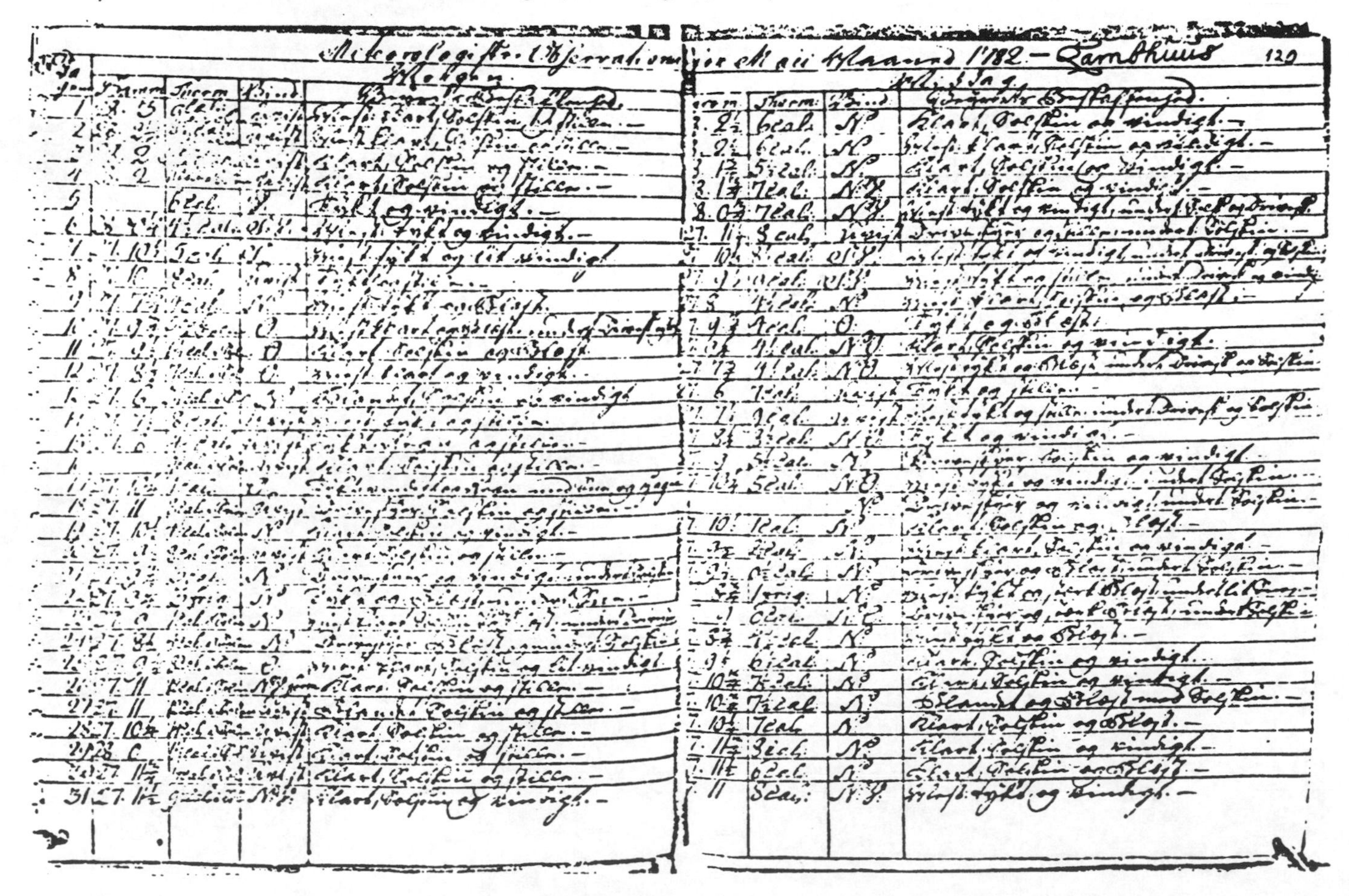

Figure 2.8. Extract from the logbook of HM Cutter *Cockatrice* from Saturday 27 December 1783 to Thursday 8 January 1784 whilst cruising off the coast of Sussex (from Oliver and Kington, 1970).

Date	Winds	Bearings & Distances at Noon	Remarks
[illegible]	[illegible]	Brighthelmston [illegible] Miles	[illegible]
[illegible]	[illegible]	At Anchor off Shoreham	[illegible]
[illegible]	[illegible]	Brighthelmston [illegible] Miles	[illegible]
[illegible]	[illegible]	Shoreham [illegible] Miles	[illegible]
[illegible]	[illegible]	Shoreham [illegible] Miles	[illegible]
[illegible]	[illegible]	Beachy head [illegible] Miles	[illegible]
[illegible]	[illegible]	Beachy head [illegible] Leagues	[illegible]
[illegible]	[illegible]	Shoreham [illegible] Miles	[illegible]
[illegible]	[illegible]	Shoreham [illegible] Miles	[illegible]
[illegible]	[illegible]	At Anchor in Shoreham Road	[illegible]
[illegible]	[illegible]	Do.	[illegible]
[illegible]	[illegible]	Shoreham [illegible] Miles	[illegible]
[illegible]	[illegible]	Shoreham [illegible] Leagues	[illegible]

3

Missing links

We have now reached the point in our story when a vast number of meteorological observations were being made and registers of the weather kept in many places in Europe with a view to discover some empirical rules to determine the changes of the weather. The supposition was that these changes obeyed some regular, though as yet unknown, laws. However, whilst significant advances had been made in the study of other natural sciences, following the collection and analysis of observations, similar progress was not so readily forthcoming in meteorology. Nevertheless, despite the inability to make any major impact on understanding, let alone predicting, day-to-day variations of the weather, it was still believed that a sustained programme of systematic observing would eventually bring to light regularities in atmospheric behaviour.

The difficulty was that although many of the daily weather reports made in Europe during the 1780s were periodically examined at the collecting centres of Mannheim and Paris, effective methods of transmitting and co-ordinating the data so as to obtain meaningful representations of current weather conditions had yet to be developed. Nonetheless, several notable investigations into the nature of storms had indicated the way in which future advances were to be made. For instance, in 1704 Daniel Defoe suggested in his essay *The Storm* that storms of wind and rain might be integral parts of migratory weather systems. In this work he made a detailed study of the disastrous storm which affected the British Isles on 26–27 November 1703 (Old Style) which, he thought, had originated near the south coast of North America, had travelled eastward across the Atlantic to affect England, Denmark and the Baltic, and finally disappeared in the Arctic region. In 1743, Benjamin Franklin began thinking along similar lines as a result of simultaneous weather observations made during a lunar eclipse by himself in Philadelphia and his brother in Boston. From these reports Franklin found that a storm may move in the opposite direction to that of the rain-bearing winds. In this case the rain blew from the north-east and yet the storm, later traced hour by hour through correspondence, had moved progressively north-eastward from Georgia to New England as a general weather system. Another pioneering investigation of atmospheric behaviour was made in the 1780s when scientists of the *Societas Meteorologica Palatina*, using observations collected at the Mannheim centre, showed that barometric minima generally moved from west to east over Europe.

Interestingly enough in themselves, these individual studies needed to be co-ordinated into a general theory of the relationships between the various meteorological elements before daily instrumental observations, such as those made in the 1780s, could be analysed effectively. After such promising beginnings the process was very slow to develop. Did 1816, the 'year without a summer' with its exceptionally cold weather and disastrous harvests, provide the stimulus for an increased effort to understand and predict weather changes? In any event the idea of mapping over a large area simultaneous daily observations of meteorological elements such as pressure, wind and temperature, the concept upon which synoptic weather studies are based, was advocated at this time by the German meteorologist Heinrich Brandes. Towards the close of 1816 he wrote:

> . . . If one could collect very accurate meteorological observations, even if only for the whole of Europe, it would surely yield very instructive results. If one could prepare weather maps of Europe for each of the 365 days of the year, then it would be possible to determine, for instance, the boundary of the great rain-bearing clouds, which in July [1816] covered the whole of Germany and France; it would show whether this limit gradually shifted farther towards the north or whether fresh thunderstorms suddenly

formed over several degrees of longitude and latitude and spread over entire countries . . . In order to initiate a representation according to this idea, one must have observations from 40 to 50 places scattered from the Pyrenees to the Urals. Although this would still leave very many points uncertain, yet by this procedure, something would be achieved, which up to now is completely new.

Breslau, 1 December 1816 (Schneider-Carius, 1955).

As a meteorological observation network did not exist at this time Brandes made use of data collected in the 1780s by the *Societas Meteorologica Palatina.* The first observations to be studied by the synoptic method were those of 6 March 1783, a day on which there had been a severe storm over western and central Europe. On a map of Europe Brandes plotted the variations from the normal air pressure on that day for each station. By connecting similar variations with lines, in effect isobars, he discovered from the distribution of these variations that pressure on that day was generally low over the area, with the barometric minimum being centred over the southern North Sea, where the barometer read over 17 Paris lines, i.e. about 51 millibars, below average (see Figure 3.1). By plotting simultaneous observations of wind direction he clearly discerned, for the first time, the relation between the distribution of pressure and wind flow and deduced the principle, rediscovered about 30 years later by the Dutch meteorologist Christophorus Buys Ballot:

Standing with your back to the wind in the Northern Hemisphere pressure will be lower on your left than on your right. The reverse is true in the Southern Hemisphere.

This remains one of the fundamental laws of synoptic meteorology.

Later, Brandes studied the history of the weather of 1783 using the synoptic method. In 1820 he published the results of this study in his *Beiträge zur Witterungskunde* (Contributions to Meteorology) but it was not until several decades later that his method of synoptic weather mapping and analysis was put into general practice. This was because it had to await the invention of the electromagnetic telegraph, which by about 1845 had been developed into a workable system for rapid communications. The proposal to use telegraphic weather networks was first made in the early 1850s, when it became possible at last to collect observations from a group of meteorological stations rapidly enough to allow weather maps to be plotted and analysed at a central office, so that forecasts could be made and issued within a time span shorter than the validity of the predictions.

However, during the late 1850s the English scientist Francis Galton became concerned that the large number of observations being collected in Europe by the newly-established national meteorological services were not being synoptically combined to give a general view of the weather over the region as a whole. As an

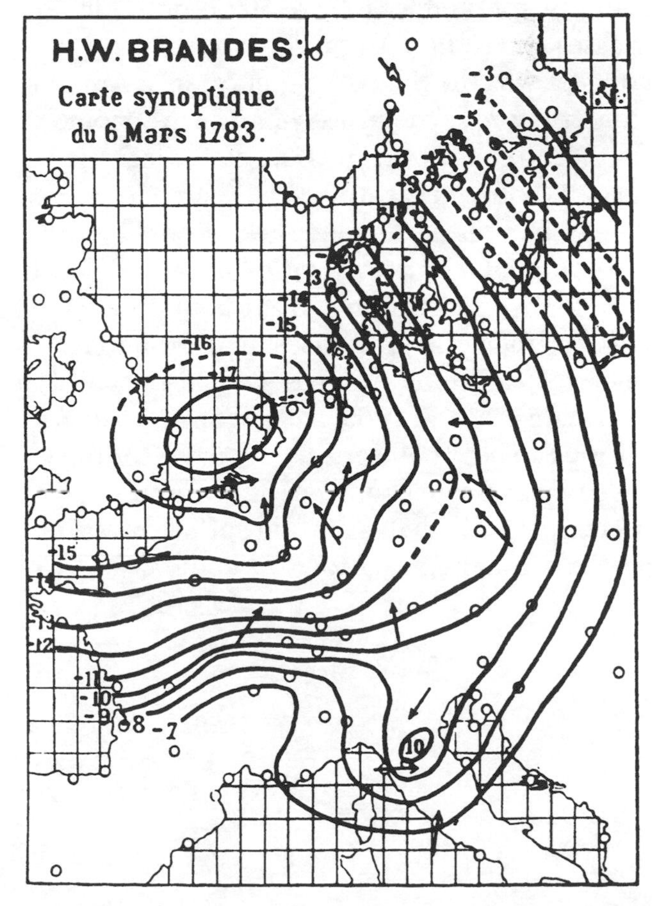

Figure 3.1. The first synoptic presentation of meteorological observations in relation to the pressure distribution as devised by Heinrich Brandes in 1820. This chart was reconstructed and published later in the nineteenth century by the Swedish meteorologist Hildebrand Hildebrandsson according to Brandes' original scheme (from Ludlam, 1966).

example of what he believed could be achieved in co-operative weather mapping Galton invited meteorologists in the British Isles and on the Continent to send him their daily reports for a trial period of a month. In 1863 he published the results of this undertaking in a work entitled *Meteorographica or methods of mapping the weather*.

This was an important contribution to synoptic meteorology. Besides presenting a series of morning, afternoon and evening weather charts with a network of observations covering Europe for each day in December 1861, Galton also included a set of smaller maps on which he depicted, for the first time, atmospheric convergence and divergence by means of flow-lines. Analysis of these latter charts enabled him to recognise the existence of high-pressure systems, or 'anticyclones' as he called them; up to that time the emphasis in synoptic studies had been on investigation of storms and low-pressure systems.

Another important ingredient which needed to be discovered before weather conditions on surface charts could be fully analysed using the synoptic method was the recognition of air masses and fronts. An early advocate of this concept was the English meteorologist Admiral Robert Fitzroy who in 1854 had been appointed first Head of the official UK meteorological service. Following a study of the synoptic charts prepared in his department he presented, in 1860, the essential features of frontal depressions which he visualised as forming on the boundary between tropical and polar air masses. Unfortunately, these forward-looking ideas were neglected after his tragic death in 1865, and synoptic practices during the next half century or so were mostly based on more orthodox but static methods of meteorological analysis using surface pressure patterns and their relation to weather.

Important advances in synoptic meteorology and weather mapping were next made during the First World War by Vilhelm and Jacob Bjerknes, father and son, co-founders of the Bergen School of Meteorology. Because of the hostilities, neutral Norway had become cut off from meteorological data and weather forecasting was severely curtailed. But this was the spur for action: under the Bjerkneses, a dense network of observing stations was established by the Norwegian Weather Service. The synoptic analysis of the detailed weather maps constructed at Bergen starting in 1918 showed that there were usually well-marked discontinuities in the wind flow and distribution of pressure, temperature and humidity around depressions. These features, later called warm and cold fronts, were found to divide low-pressure systems into warm and cold sectors containing tropical and polar air masses respectively. The three-dimensional model of a depression developed by the Bergen School, with its warm and cold fronts and related cloud and precipitation distributions, make it possible to integrate meteorological observations covering extensive areas into clearly definable weather systems, whose movement and development can then be related directly to the general circulation.

These steps in the advancement of synoptic meteorology from Brandes to the Bjerkneses illustrate the advantage of working with hindsight: one is able to apply concepts discovered, formulated and developed during the nineteenth and twentieth centuries to historical weather observations made in the eighteenth century. It is therefore particularly gratifying that so many good quantitative instrumental observations were made in the 1780s, that they have survived to the present day, and that we now have the means and opportunity to apply present synoptic methods and techniques to virtually untapped sources of daily weather data.

4

A bi-centenary exercise

Two questions come to mind when considering daily weather mapping in the historical-instrumental period. How similar are the methods employed to those used in the plotting and analysis of current synoptic charts? And what differences in the methods arise from the historical nature of the observations? A discussion of the actual procedure adopted in the preparation of the daily weather maps for the 1780s will highlight these similarities and differences.

In present-day synoptic weather mapping a number of charts are plotted and analysed at forecast offices every day, with those for the main synoptic hours 0001, 0600, 1200 and 1800 GMT being the most widely used. At the commencement of this project it was decided that although observations in the 1780s were made at many of the stations two or three times a day, only one chart a day would normally be drawn, with 1400 h being selected as the best time, for a number of reasons.

It is known from meteorological theory that if the effect of surface friction is ignored, the geostrophic wind, which approximates to the actual wind at a height of about one kilometre (3300 feet), can be considered to blow parallel to the isobars on mean sea-level pressure charts. Lowest pressure is on the left, and the speed of the wind is inversely proportional to the isobar spacing, i.e. close isobars, strong wind; widely-spaced isobars, light wind.

Over land, the surface wind is most-closely related to the geostrophic wind and hence mean sea-level pressure patterns at about 1400 h to 1500 h, the time of maximum heating, when vertical mixing in the friction layer is at its strongest. At that time the surface wind will be brought nearer to the geostrophic wind, both in speed and direction, because the interchange of surface and upper air in the friction layer is at its maximum. Surface winds are then generally about one-third of the geostrophic speed and backed in direction across the isobars by about 30 °. Overnight, however, increasing stability in the friction layer usually decreases vertical mixing, resulting in less exchange between surface and upper levels, and surface winds, with the same pressure gradient, are slower in speed and more backed in direction than during the day. The ratios of speeds and amounts of backing vary in both time and space, and in extreme cases the surface wind may show no relation at all to the geostrophic wind. Thus estimations of the alignment and spacing of surface isobars from land stations reporting wind velocities but not pressure values would be more problematical on charts drawn up say at 0700 h or 2100 h rather than at 1400 h.

Over the sea, other things being equal, there is little diurnal variation in vertical mixing because there is little change in surface temperature, and the relation of surface wind to the geostrophic wind is essentially unaffected by the time of day. With less frictional drag over the sea, surface winds have speeds of about two-thirds of the geostrophic value and are backed in direction by only 10 ° to 15 ° across the isobars, with little change occurring by day or night.

Taking these differences into account, the concept of the geostrophic wind is a useful aid in weather mapping. Using it, observations of wind velocity can be quantitatively combined within and to some extent beyond the pressure data network.

There are three further advantages in drawing daily weather maps at 1400 h. First, it seems to have been common practice amongst many eighteenth-century meteorological observers to record the state of the weather, even if only once a day, at midday or during the early afternoon. Second, as maximum temperature is a relatively conservative air mass property, any significant rise or fall of the temperature at 1400 h compared to the previous day would probably indicate that a new air mass had been advected over the station concerned during the past 24 hours. Third, for the purpose of classifying weather types and circulation patterns, it is convenient to have a chart drawn at about midday to represent the general synoptic weather situation for each 24-hour period.

Figure 4.1 shows the synoptic coverage available for the 1780s. The stations plotted are divided into three main groups: those belonging to the two main scientific societies on the Continent making concerted efforts to collect daily instrumental meteorological data; those made by private individuals; and lastly, locations in sea areas and coastal regions typically representing the data obtained from ships' logbooks.

The collection of data has involved a wide-ranging enquiry and search operation with record offices, archives, libraries and private sources throughout Europe, and is, of course, a continuing process. When sufficient weather data for a given period of time has been collected, the next task is to reduce the observations to a standard format for plotting on the charts. This is a lengthy and exacting process since in contrast to the internationally agreed procedure of today, not only is one dealing with data in several different languages, sometimes in the form of original handwritten manuscripts, but also with a variety of terms and units, all of which have to be reduced to present-day equivalents.

In order to make an effective inroad into the mammoth task of data reduction and plotting, a start is usually made with the systematic group of observations of the *Societas Meteorologica Palatina* extracted from the appropriate volume of the Society's *Ephemerides.* The synoptic coverage is gradually built up, station by station, working with a period of a month at a time. Daily reports from about 20 to 30 stations are available from the Palatinate source throughout the 1780s. The material is clearly set out in printed form but it still has to be examined for conversion of units, terms and symbols into present-day equivalents. This organisation was making great advances in standardising meteorological readings and observational practice, using scientific Latin to overcome language differences. Nevertheless, there were a few regional variations within the system; for example, stations in Russia recorded temperature in degrees Delisle instead of on the standard scale of Réaumur. On the Delisle scale the boiling point of water was taken as zero and the scale inverted so that low temperatures were expressed by high positive numbers. The freezing point of water was defined as 150 ° and the scale extended to 240 ° or 270 ° (−60 °C to −80 °C), so as to record the extremely low temperatures experienced at times in Moscow and St Petersburg (Leningrad) during winter.

Having converted the instrumental readings into current units, the descriptive terms used to define wind strength, state of the sky, significant weather, and optical and atmospheric phenomena are then examined for transformation into present plotting symbols, using the information given in Tables 2.2, 4.1, 4.2 and 4.3. The observations of the stations in the Palatinate network are then ready for plotting.

As an example let us take the observation made by Johann Hemmer at 1400 h on 9 January 1785 at Mannheim (see Figure 2.5, p. 13).

Barometer: 28 Paris inches, 2.8 lines. This is converted to millibars (mb) and reduced to Mean Sea Level (MSL) pressure to give a value of 1033 mb.

Thermometer interior: 4.3 ° Réaumur (R). (Not plotted)

Thermometer exterior: 1.8 ° R. This is converted to degrees Celsius (C), i.e. 2 °C.

Hygrometer: 20.4 ° (Not plotted)

Although humidity readings were taken, the instrument had not yet been perfected and the scale remains obscure. The *Societas Meteorologica Palatina* recognised this deficiency and a prize was offered for a hygrometer with a stable sensitivity that could be properly calibrated.

Magnetic declination: 19 °48′ (Not plotted)
Wind: NE, 1.
Rainfall: 151 p.lin. 64ths (Not plotted)
Evaporation: (Not plotted)
Height of Rhine: −3 ft 6 ins. (Not plotted)
Position of Moon by zodiacal sign: (Not plotted)
State of the sky: 0700 h: Total cloud cover;
1400 h: Mostly cloudy;
2100 h: Clear sky.

Figure 4.1. Map of stations showing the synoptic coverage available for the 1780s.

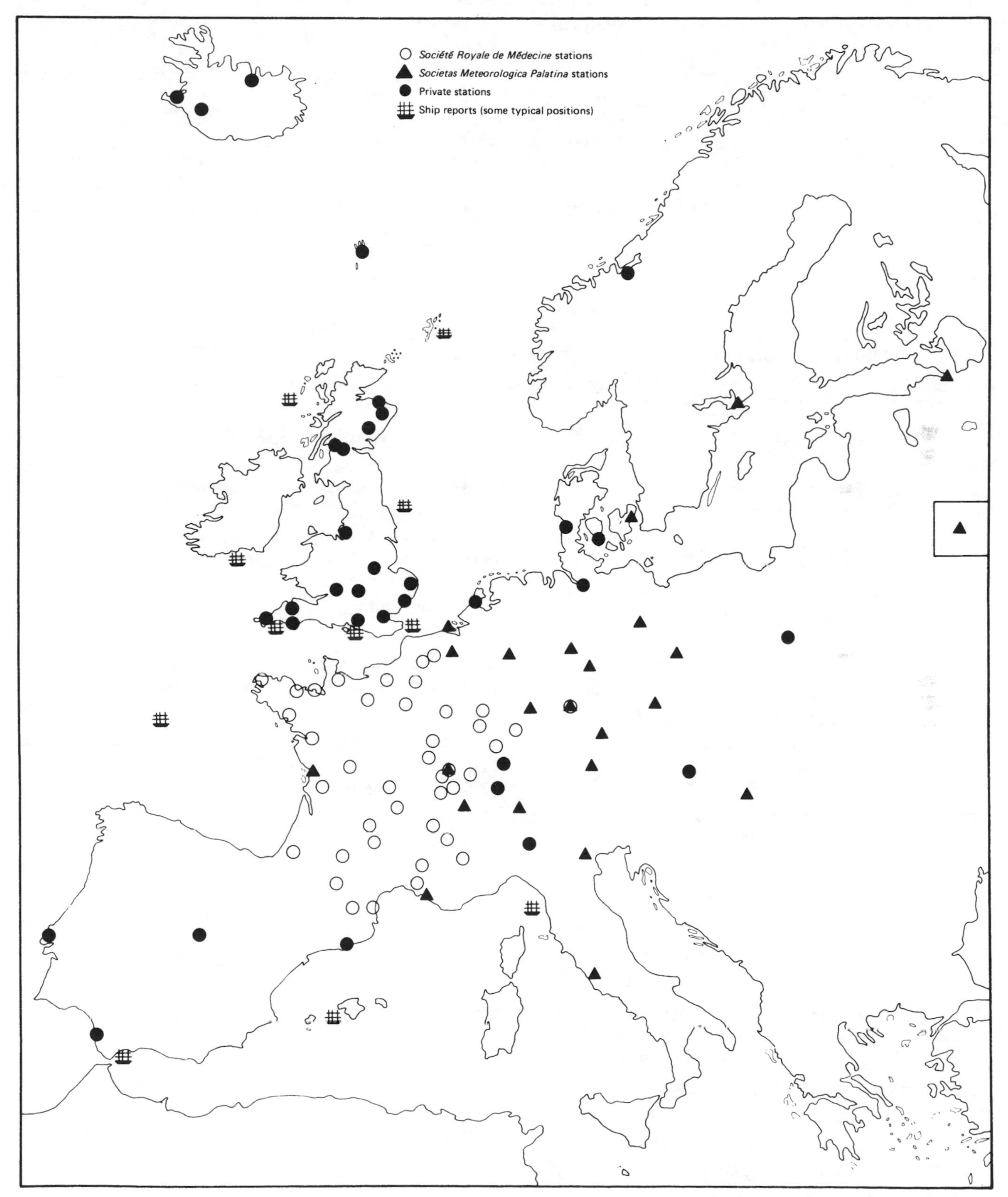

Table 4.1 Societas Meteorologica Palatina *definitions and plotting symbols for state of the sky.*

Coeli facies/State of sky		
SMP symbol	Definition	Plotting symbol
⊙	*Ex omni parti serenum quo tamen in statu si pallidor solis stellarumque lux/* Clear sky	○
⊙+	*Feurit, huic sign crux adjungitor/* Hazy sky, pale or weak sunlight or stars	○
-8-	*Nubes rarissimus hinc inde dispersas/* Small amounts of clouds, dispersing	⊖
-=	*Nubes variores, quae minorem coeli partem occupant/* Less than half cloud cover	⊕
=	*Coelum nubes inter coeruleum colorem ex aequo divisum/* Partly cloudy with clear skies in between	◍
-¤-	*Nubes majori coeli parti inductas, sive continuae sint, sive disjunctae et quasi pertusae/* Mostly cloudy, either continuous or broken	◍
= =	*Totum nubibus tectum/* Total cloud cover	●

N.b. No entry in column indicates sky obscured.

Weather: Snow during preceding night; fog at 0900 h and in mountains all day.

The relevant information is now plotted around the appropriate station circle (see Figure 4.2(a)). A similar type of procedure is adopted for the observations of the *Société Royale de Médecine* network, which amount to a further 20 to 30 station plots throughout the 1780s.

The observation made by Dr Maret at 'midi' on 11 October 1781 at Dijon (see Figure 2.3, p. 6) is selected as an example of the plotting from this source.

Barometer: 27 Paris inches, 7.6 lines. This is converted to millibars and reduced to MSL pressure to give a value of 1027 mb.

Table 4.2 Societas Meteorologica Palatina *definitions and plotting symbols for significant weather.*

Meteora/Weather		
SMP symbol	Definition	Plotting symbol
¦¦	*Pluvium*/Rain	:
++ ++	*Nivem*/Snow	⁑
∷	*Grandinem*/Hail	⍒
⋰	*Pruinam*/Frost	⊔
⁖	*Nebulam*/Fog or mist	≡
♂	*Tempestatum*/Thunderstorm	⌈ζ

N.b. The supplementary symbol ✳ was used as an indication of intensity, e.g. ¦¦ ✳ heavy rain; ⁖ ✳ thick fog; ♂ ✳ severe thunderstorm.

Table 4.3. Societas Meteorologica Palatina *definitions and plotting symbols for optical and atmospheric phenomena.*

SMP symbol	Definition	Plotting symbol
⋰⋱	Rainbow	⌒
◎	Solar corona or halo	⊕
©	Lunar corona or halo	⊍
⊙--⊙	*Parhelion*/mock suns	
(- -)	*Paraselenae*/mock moons	
AB	*Aurora borealis*	⏏
☄	*Stellam cadentem*/ Shooting star	
☼	*Globum igneum*/ Meteor or fireball	

Thermometer: 10.6 °R. This is converted to degrees Celsius, i.e. 13 °C.
Wind: SE.
State of the Sky: Clear.
Weather:
Matin: Hoar frost and fog which lifted and cleared.
Midi: —
Soir: Slight rain.

The relevant information is now plotted around the appropriate station circle (see Figure 4.2(b)).

Although the plot is for 1400 h, notice is taken of the morning and evening weather if significantly different.

Incidentally, Dr Maret was a member of both the *Société Royale de Médecine* and *Societas Meteorologica Palatina*, and copies of his meteorological register are available from both sources housed at Paris and Bracknell respectively.

Having completed the observations from the two scientific societies, we now come to the reports from individual observers contained in diaries, journals and meteorological registers; these amount to another 20 or so stations. Various terms expressed in a number of different languages, including Danish, Dutch, English, German, Icelandic, Latin and Spanish, were used for describing wind strength, state of the sky and significant weather. Dealing first with those that made instrumental observations, two reports have been selected as examples of this type of material: Lyndon Hall, Rutland, and Lambhús, Iceland.

Observation made by Thomas Barker at 1455 h on 30 June, 1783 at Lyndon Hall, Rutland (see Figure 2.6, p. 15).
Barometer: 29.83 English inches. This is converted to millibars and reduced to MSL pressure to give a value of 1028 mb.
Thermometer: House (interior): 68.4 ° Fahrenheit (F) (not plotted)
Abroad (exterior): 78.8 ° F. This is converted to degrees Celsius, i.e. 26 °C.
Clouds: S, 1.
Wind: NE, 1.
Weather: Cloudy, calm morning; sunny, very hot day; evening thick smoky air and smell of fens.

The relevant information is now plotted around the appropriate station circle (see Figure 4.2(c)).

By reporting the direction and speed equivalent of clouds, Thomas Barker was in fact making some pioneering observations on upper wind flow.

Observation made by Rasmús Lievog at midday on 7 May 1782 at Lambhús, Iceland (see Figure 2.7, p. 16).
Barometer: 27 Paris inches 10⅓ lines. This is converted to millibars and reduced to MSL pressure to give a value of 1016 mb.
Thermometer: 8½ °R. This is converted to degrees Celsius, i.e. 11 °C.
Wind: SW, 2.
Weather: Mostly cloudy and windy, sometimes broken skies and sunny.

The relevant information is now plotted around the appropriate station circle (see Figure 4.2(d)).

Figure 4.2. Examples of station plots.
(a) Mannheim, 9 January 1785
(b) Dijon, 11 October 1781
(c) Lyndon Hall, Rutland, 30 June 1783
(d) Lambhús, Iceland, 7 May 1782
(e) Norwich, 8 January 1784
(f) HM Cutter *Cockatrice*, off Sussex coast, 1 January 1784.

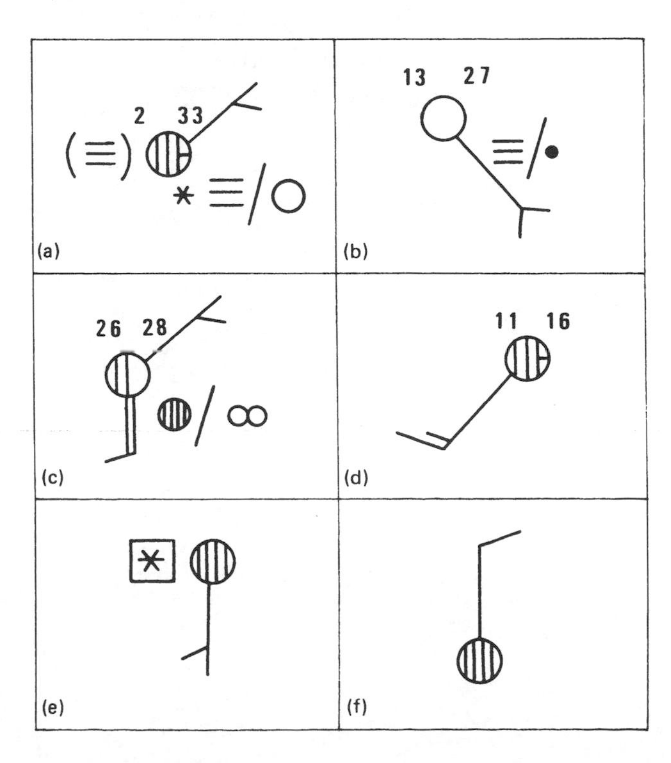

Having now reduced and plotted all the available instrumental data from private observers, next comes the turn of the purely descriptive accounts. The journal of William Youell, manager of the New Mills, in Norwich, is a typical example of this kind of weather report. His diary, as well as recording information about his milling affairs and family, also includes daily references to the weather which so often played an important role in his working operations. On 8 January 1784 he wrote:

> Did not freeze sharp last night nor today. A great number of people upon the ice on Mill Damm. Very little wind from SE, S and N. Close all day.

This information is now plotted around the appropriate station circle (see Figure 4.2(e)).

The final material to be prepared for plotting comprises the wind, weather and sea observations contained in ships' logbooks. An extract from the logbook of HM Cutter *Cockatrice* from Saturday 27 December 1783 to Thursday 8 January 1784 whilst cruising off the coast of Sussex is given in Figure 2.8 (p. 17). The entry for Thursday 1 January 1784 gives an example of the plotting from this type of source.

> Wind: ESE becoming variable becoming N.
> Remarks: First Part (1200–2000 h on previous day): fresh gales with snow and rain.
> Middle Part (2000–0400 h): fresh breezes and cloudy.
> Latter Part (0400–1200 h): light airs and cloudy.

This information is now plotted around the appropriate station circle (see Figure 4.2(f)). If significantly different, the wind and weather data in the First Part are plotted around the station circle on the previous day's chart.

This completes the plotting of the daily synoptic weather maps for the given monthly period. The next stage, the analysis of the charts, may be divided into a number of steps. First, the observations of temperature, wind, weather, and cloud plotted over each chart are examined to distinguish areas affected by different air masses. Second, a preliminary analysis is made to determine the boundaries between the air masses and hence to identify and locate any fronts. When available, variations in wind, weather and cloud reported by individual stations during the course of the day are useful in determining the passage of fronts at spot points on the chart. Also, as already mentioned earlier, variations of maximum temperature from day to day at individual stations can be indicative of air mass changes. Third, the pressure pattern is constructed by drawing isobars in accordance with the plotted pressure readings and wind reports, so as to produce a realistic air-flow pattern. As discussed at the beginning of this chapter, the direction and strength of the wind provide a valuable guide for drawing the isobars using the relation between surface and geostrophic wind flow. Isobars are drawn at 4 millibar intervals using even-numbered values, for example 996, 1000, 1004, etc. As the pressure field takes shape it can help to confirm and possibly amplify the preliminary frontal analysis. Finally, synoptic continuity in the movement and development of pressure systems and fronts is taken into account from chart to chart.

5

Daily synoptic weather maps, 1781–5

Isobars are drawn on the charts at 4 mb intervals, with the 1012 mb isobar being marked '12'. This isobar has been chosen as it usually separates anticyclonic from cyclonic systems. Occasionally however, when all the pressure values on the charts are either below or above 1012 mb, the 996 mb ('96') isobar or 1028 mb ('28') isobar are labelled instead.

The maps for each month are arranged for reading from left to right across each page, so that there are 16 days to a page. The whole month is displayed as a double-page spread.

1	2	3	4
5	6	7	8
9	10	11	12
13	14	15	16

17	18	19	20
21	22	23	24
25	26	27	28
29	30	31	

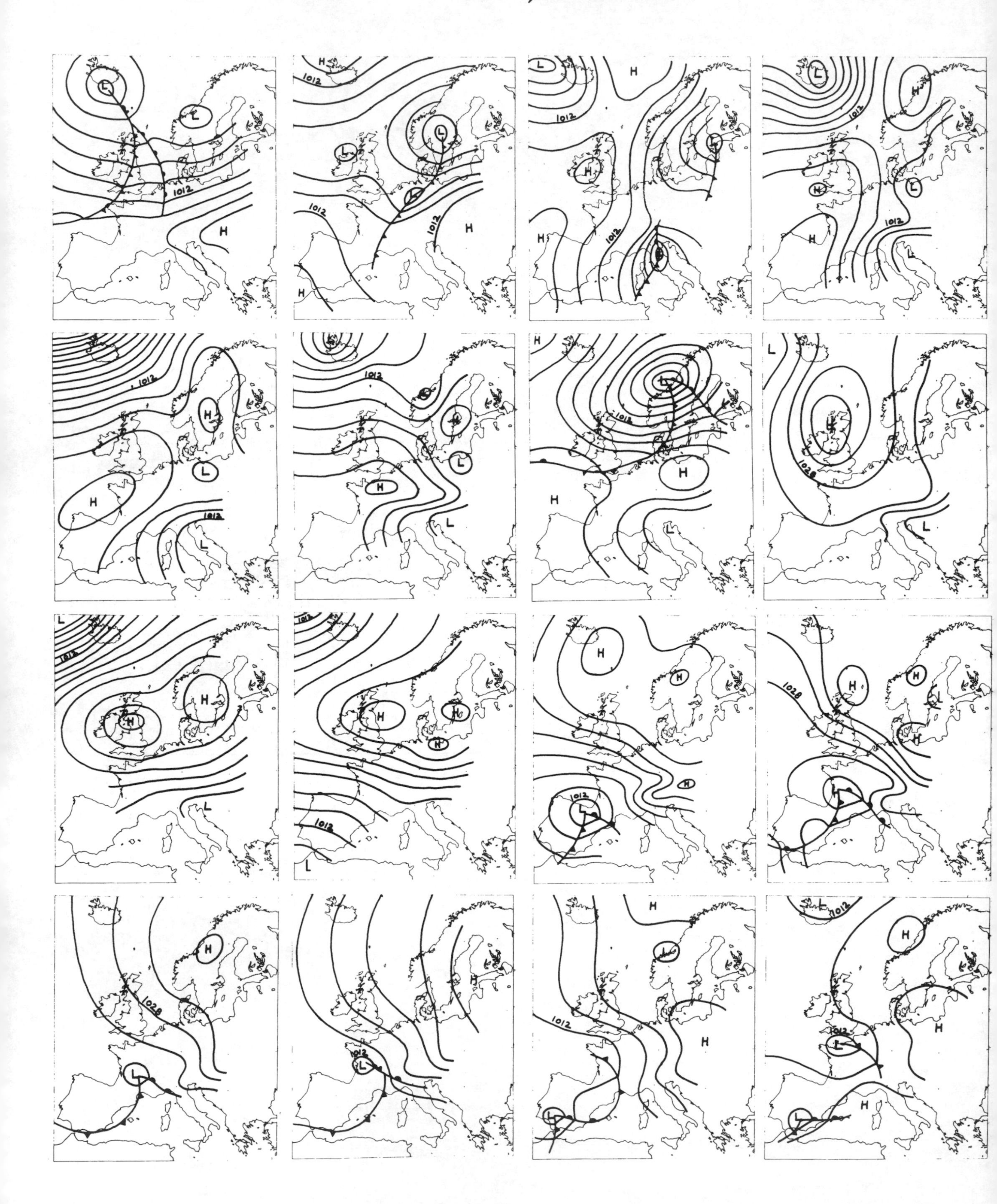
1012
1028
H
L

1012
H
L

1781 FEBRUARY

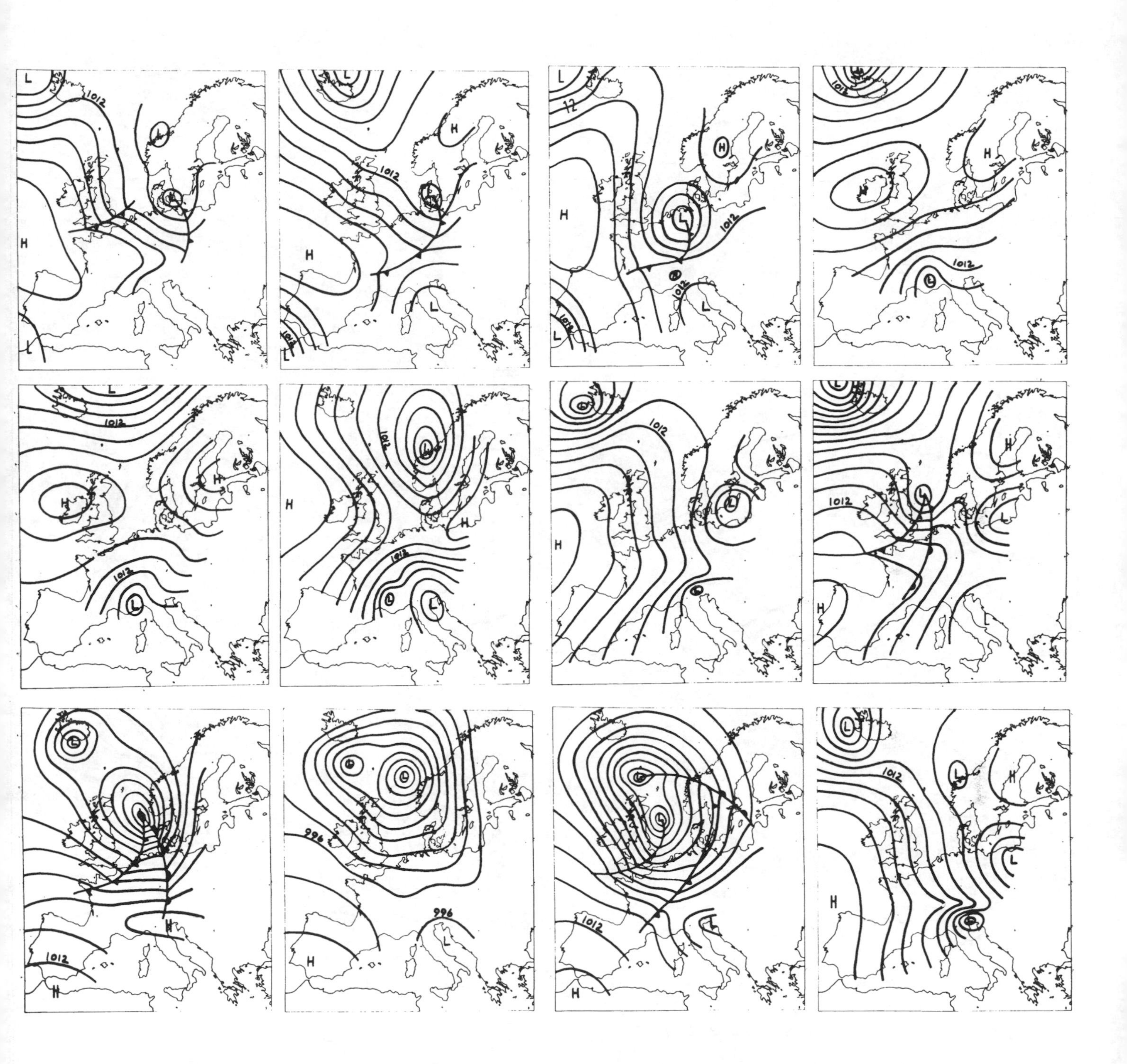

1781 MARCH

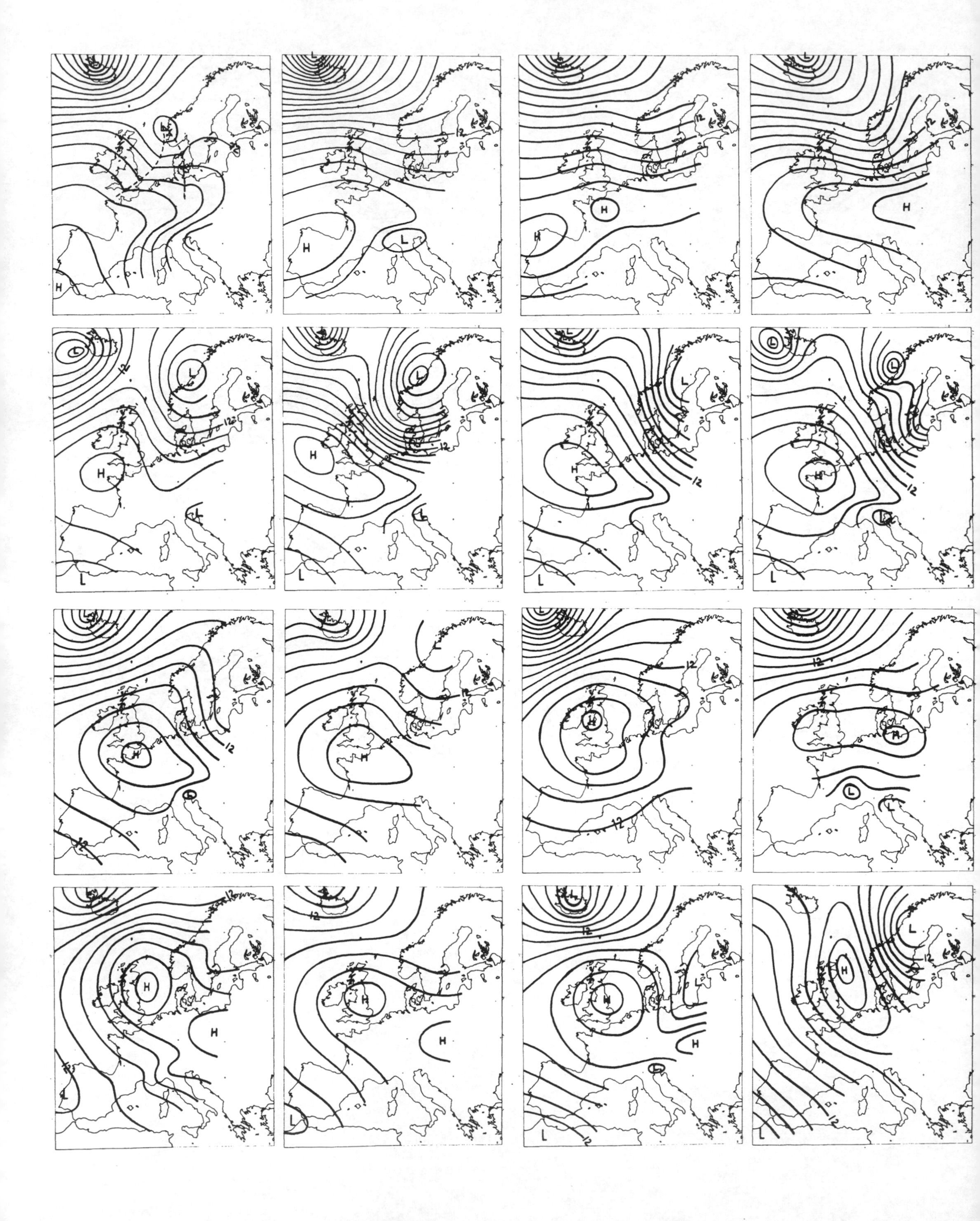

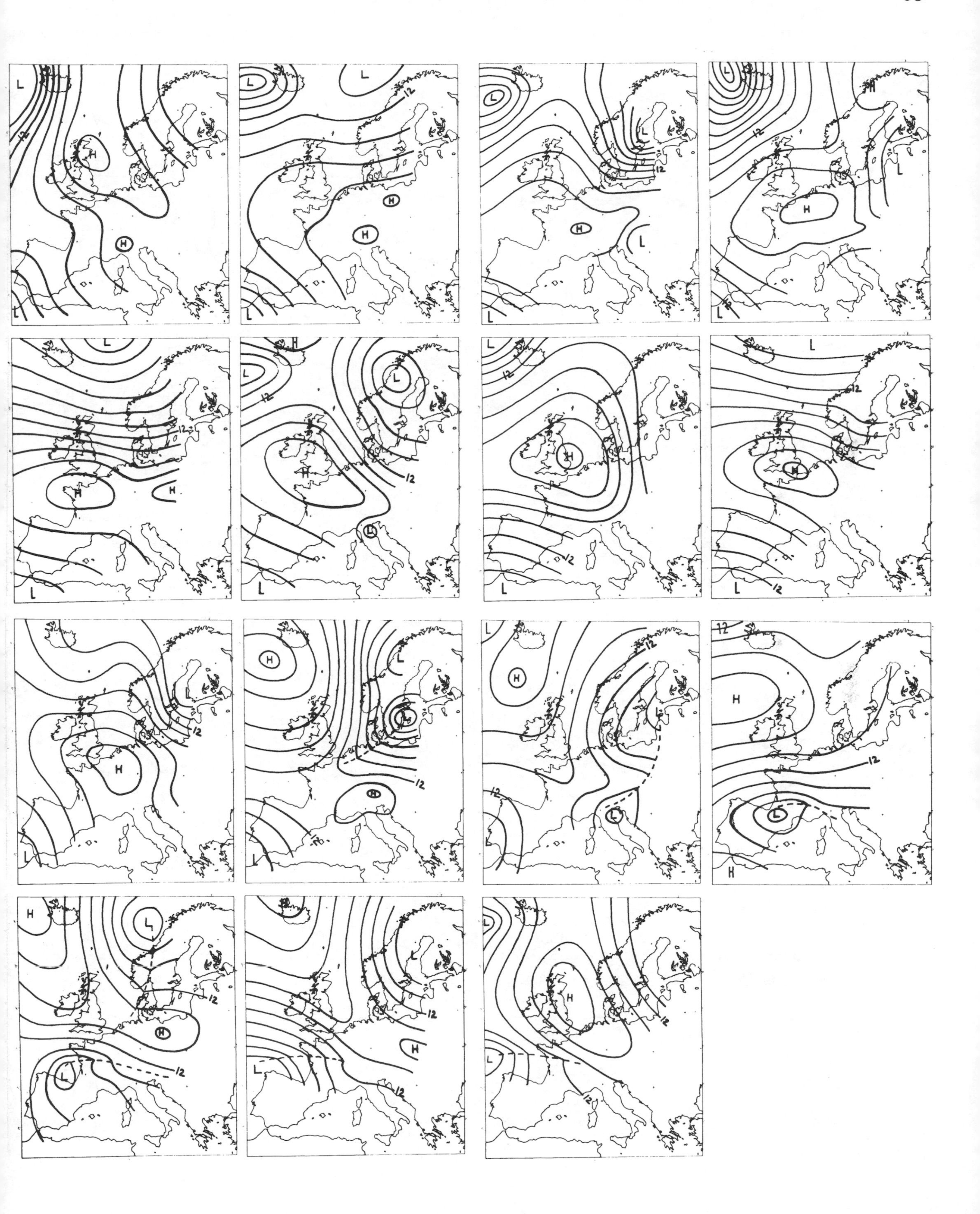

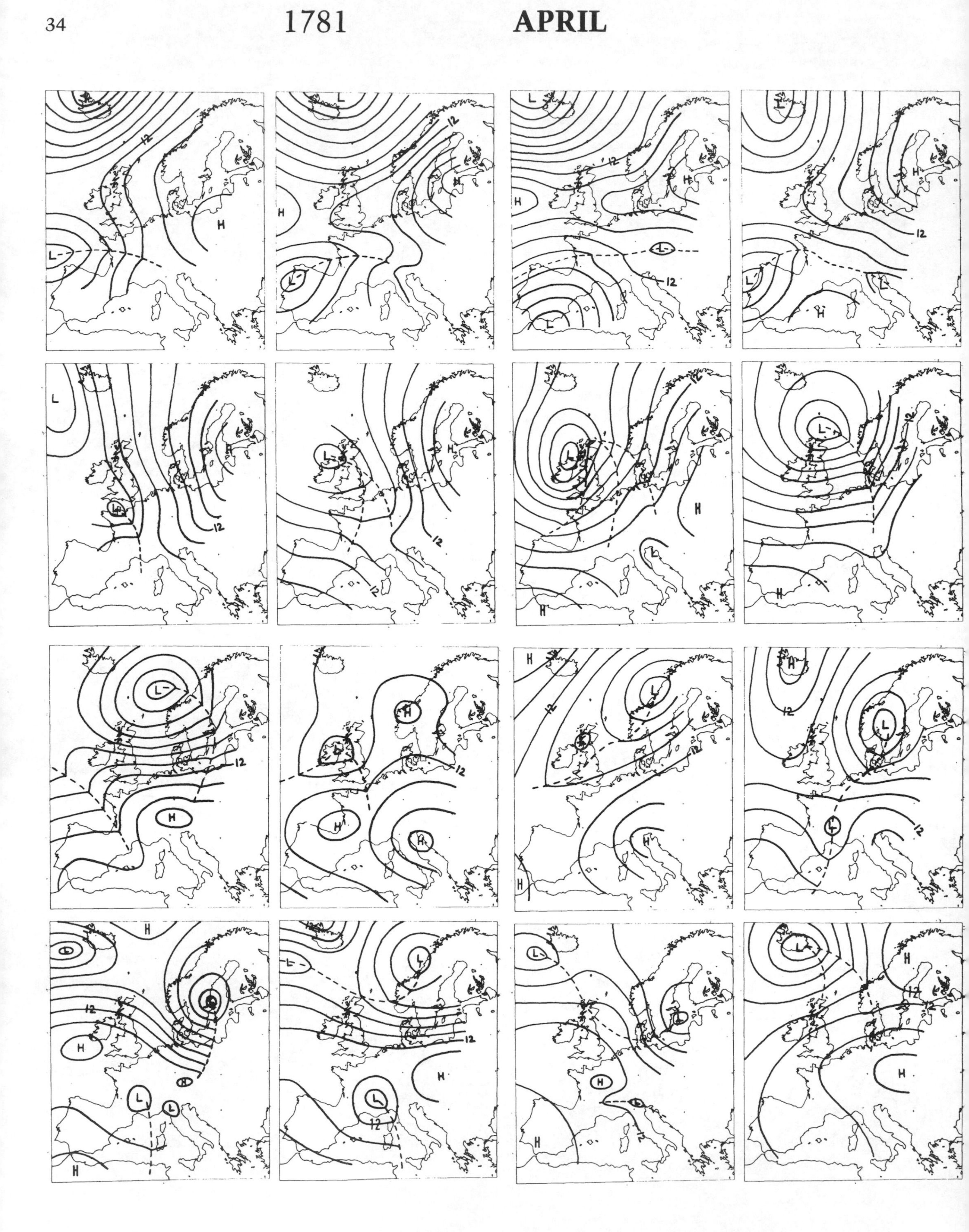

L
H
12
L
H
12
L
H
12
L
H
12

L
H
12

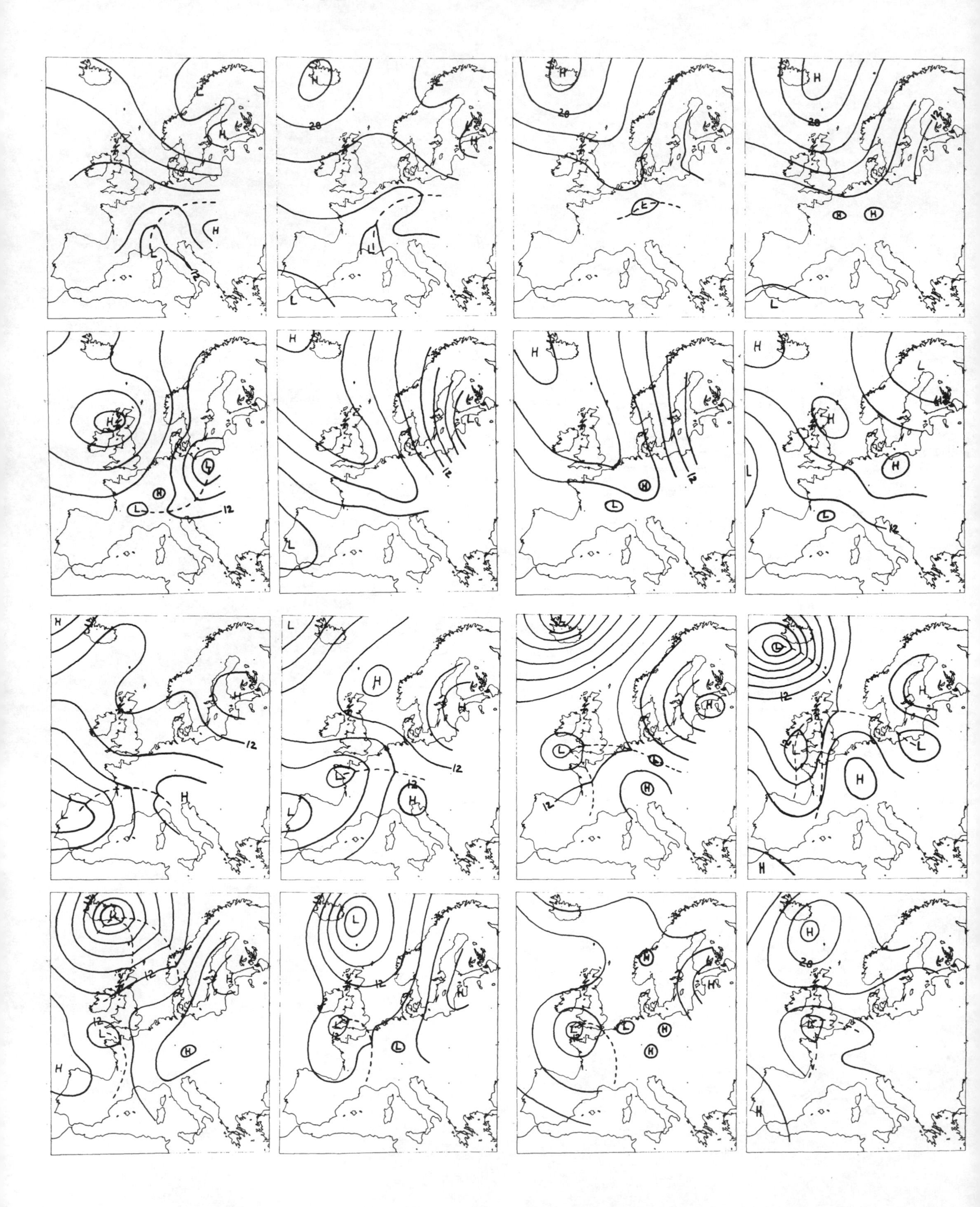

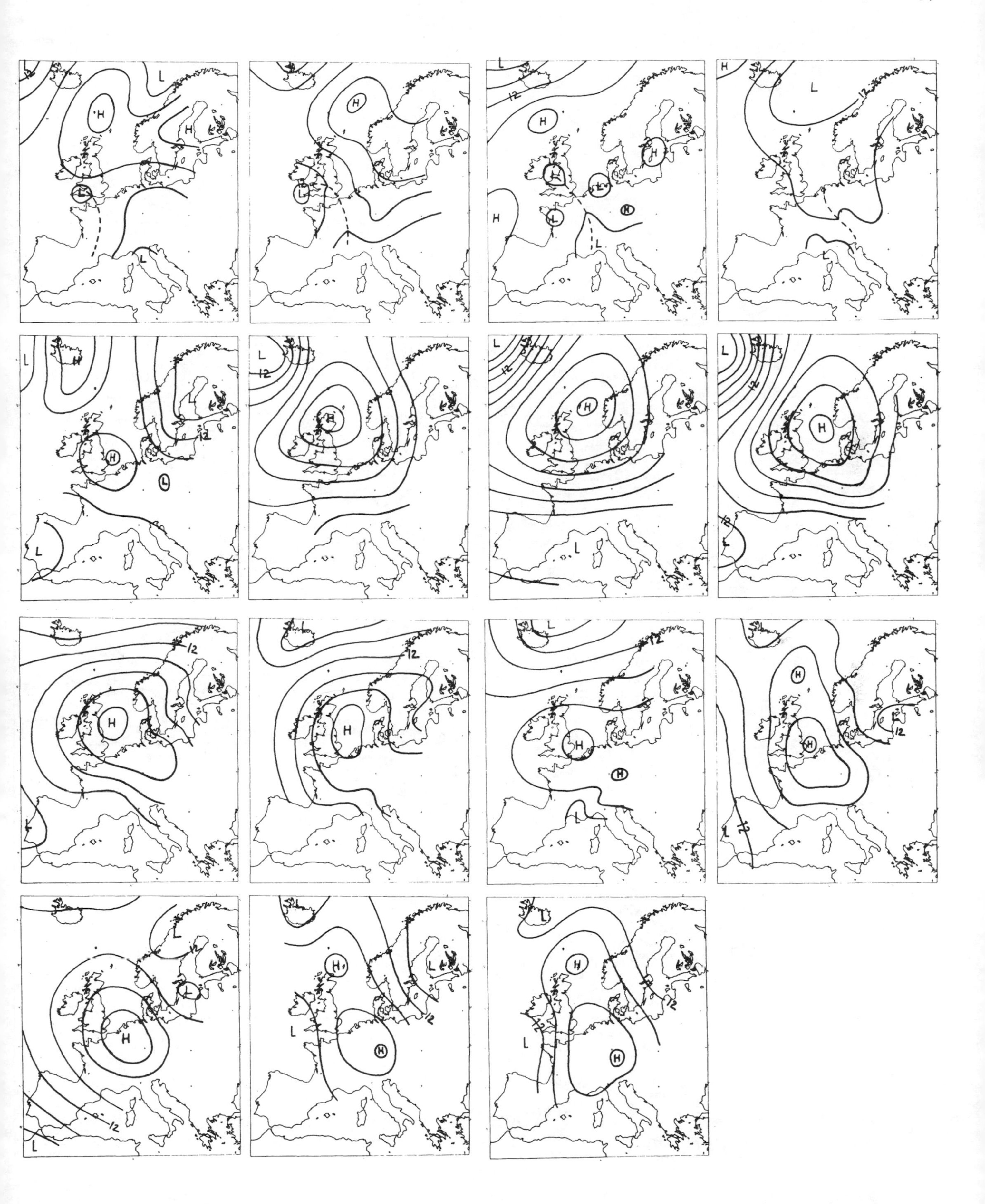
H
L
12

1781 JUNE

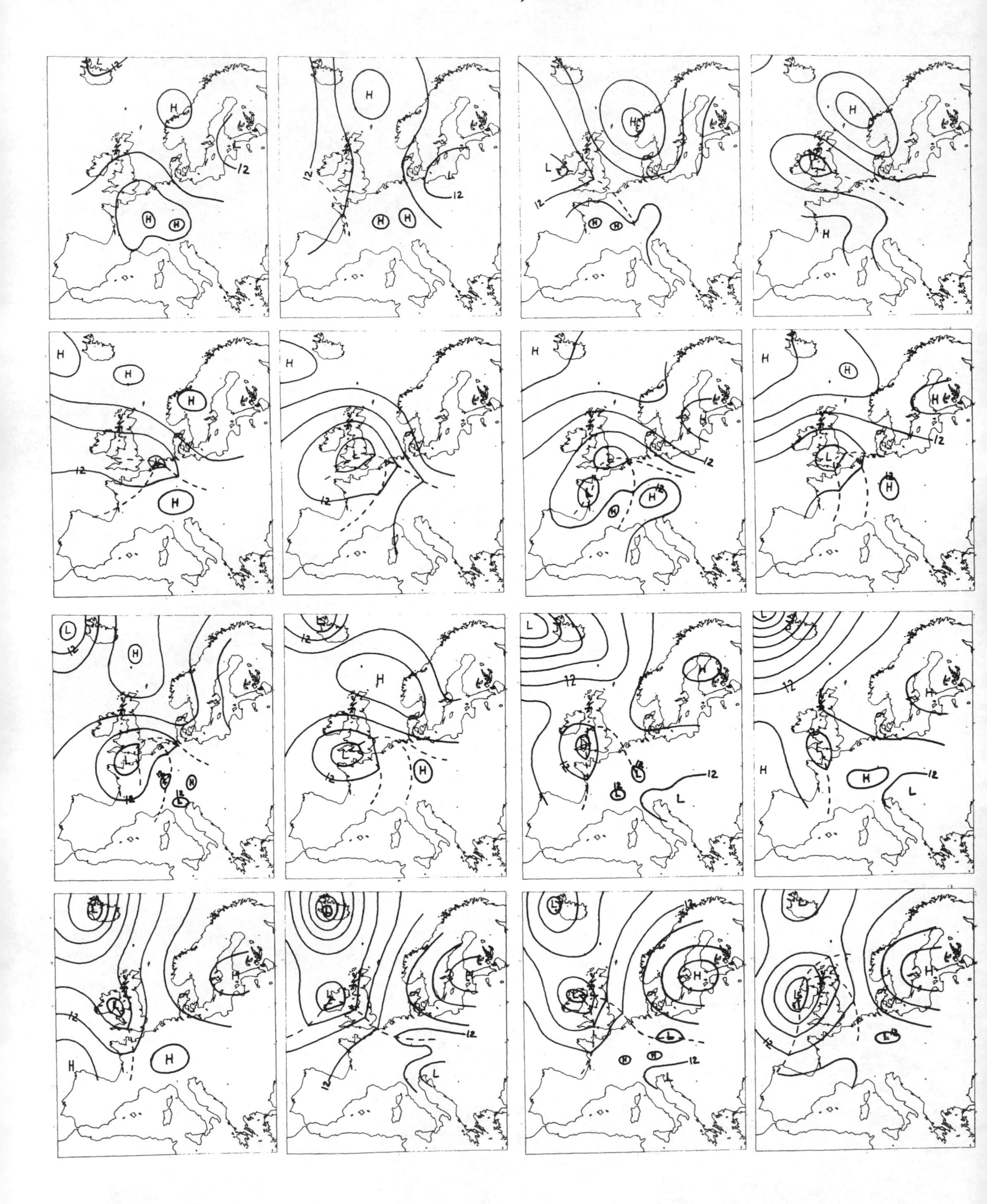

H
L
12

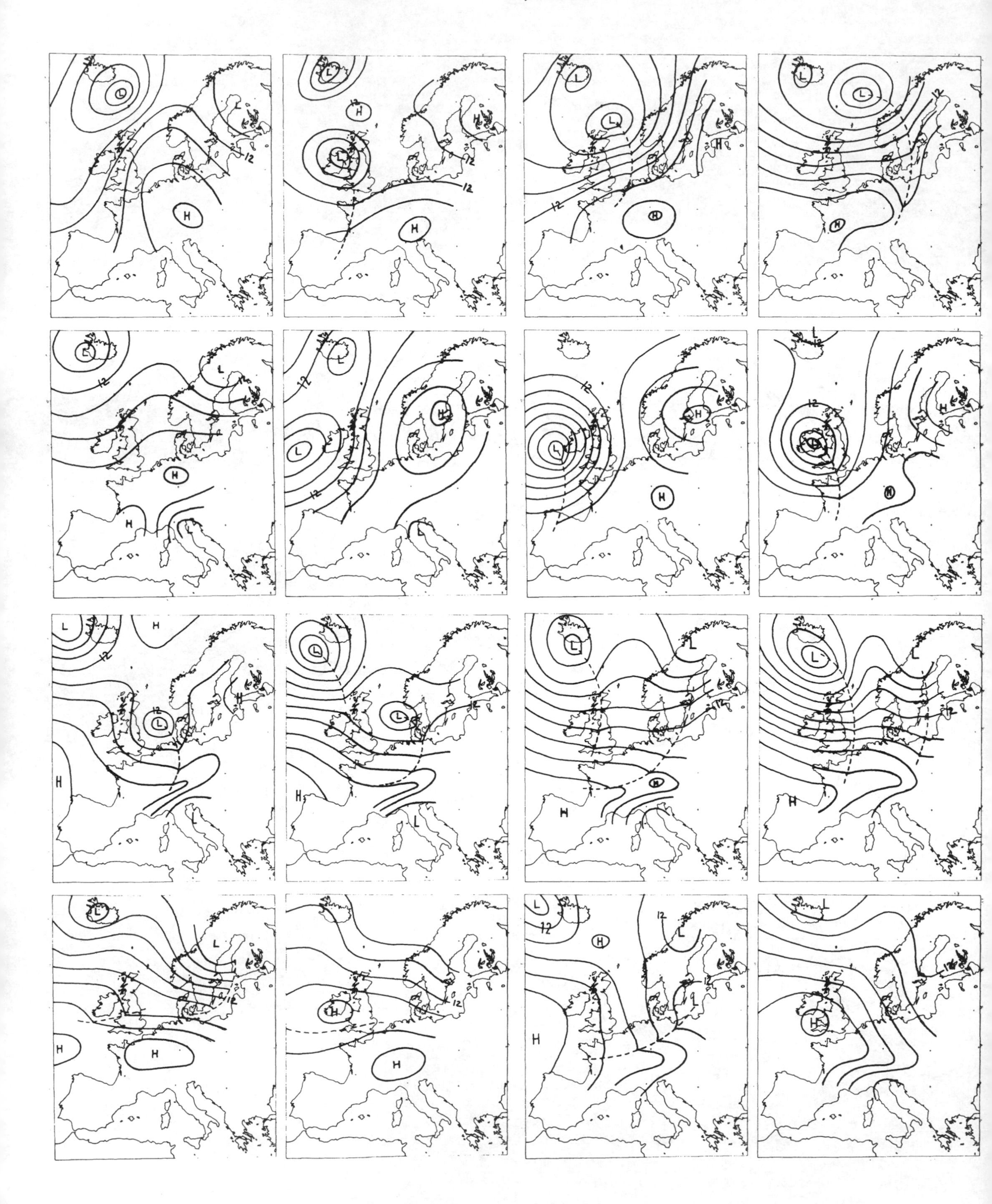

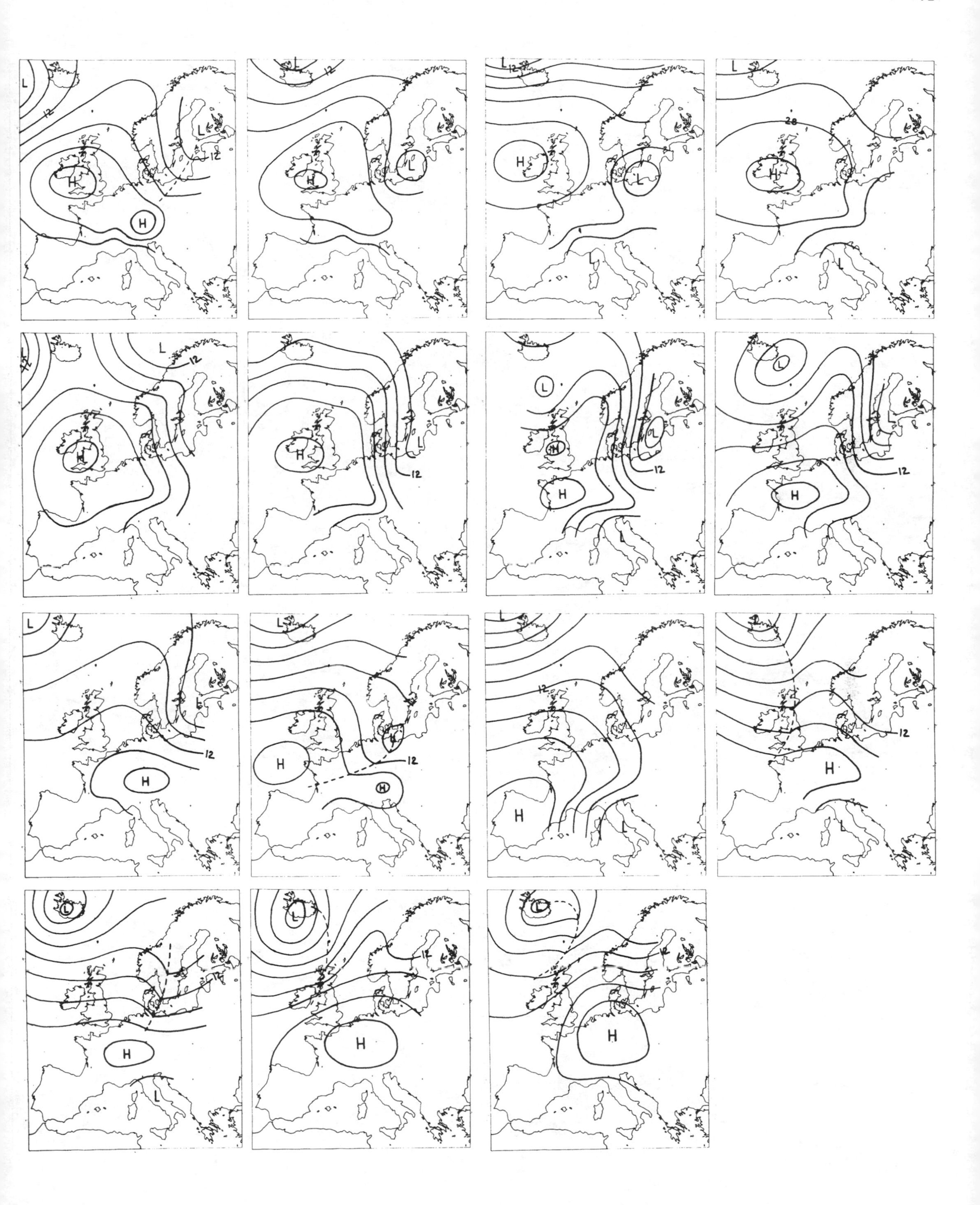

L
12
L
12
H
H
L
12
H
L
L
12
H
L
L
28
H
L
L
12
H
12
H
L
L
H
L
H
12
L
H
12
L
H
12
L
H
12
H
H
L
12
H
L
12
H
L
L
H
L
H
L
H

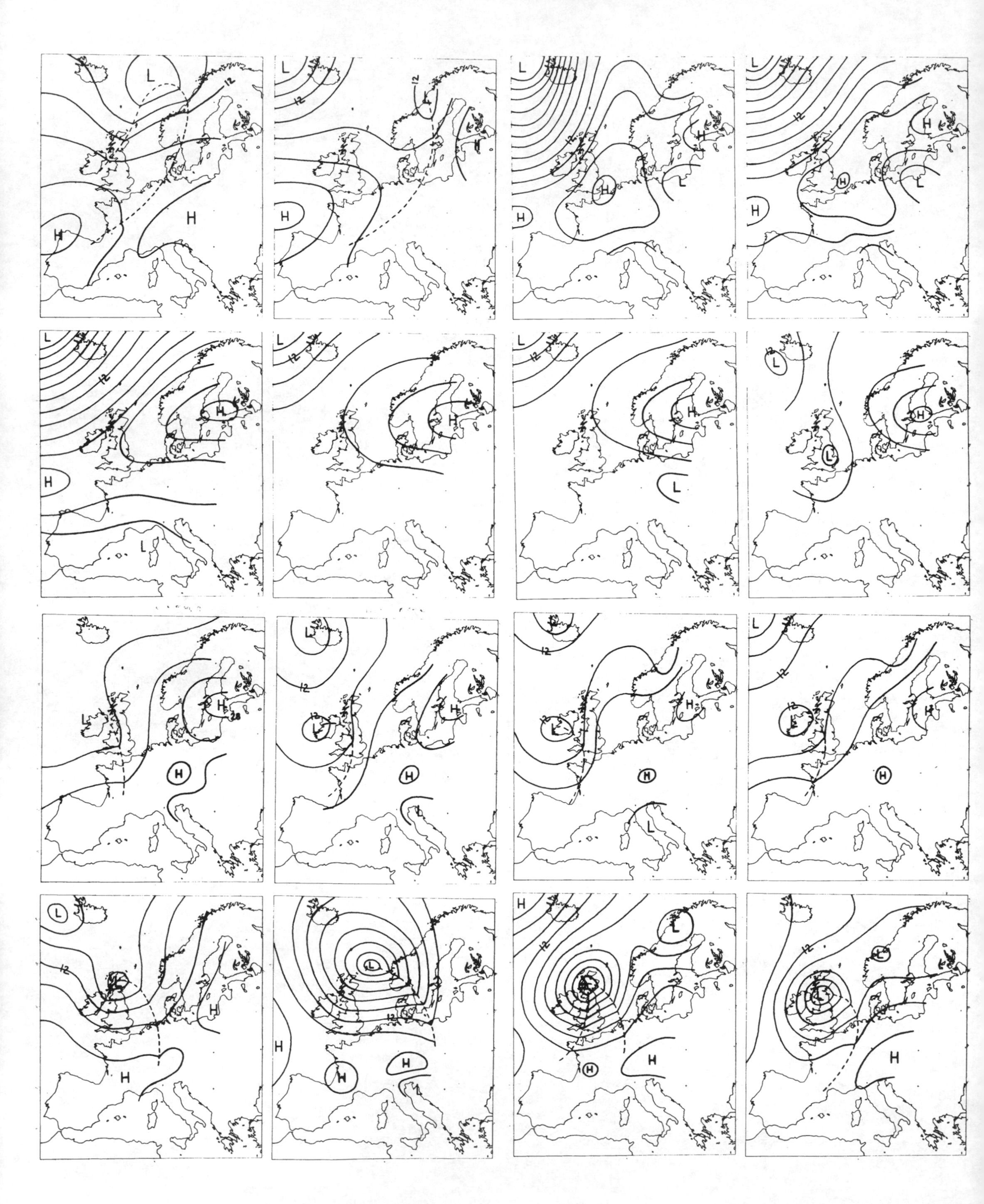

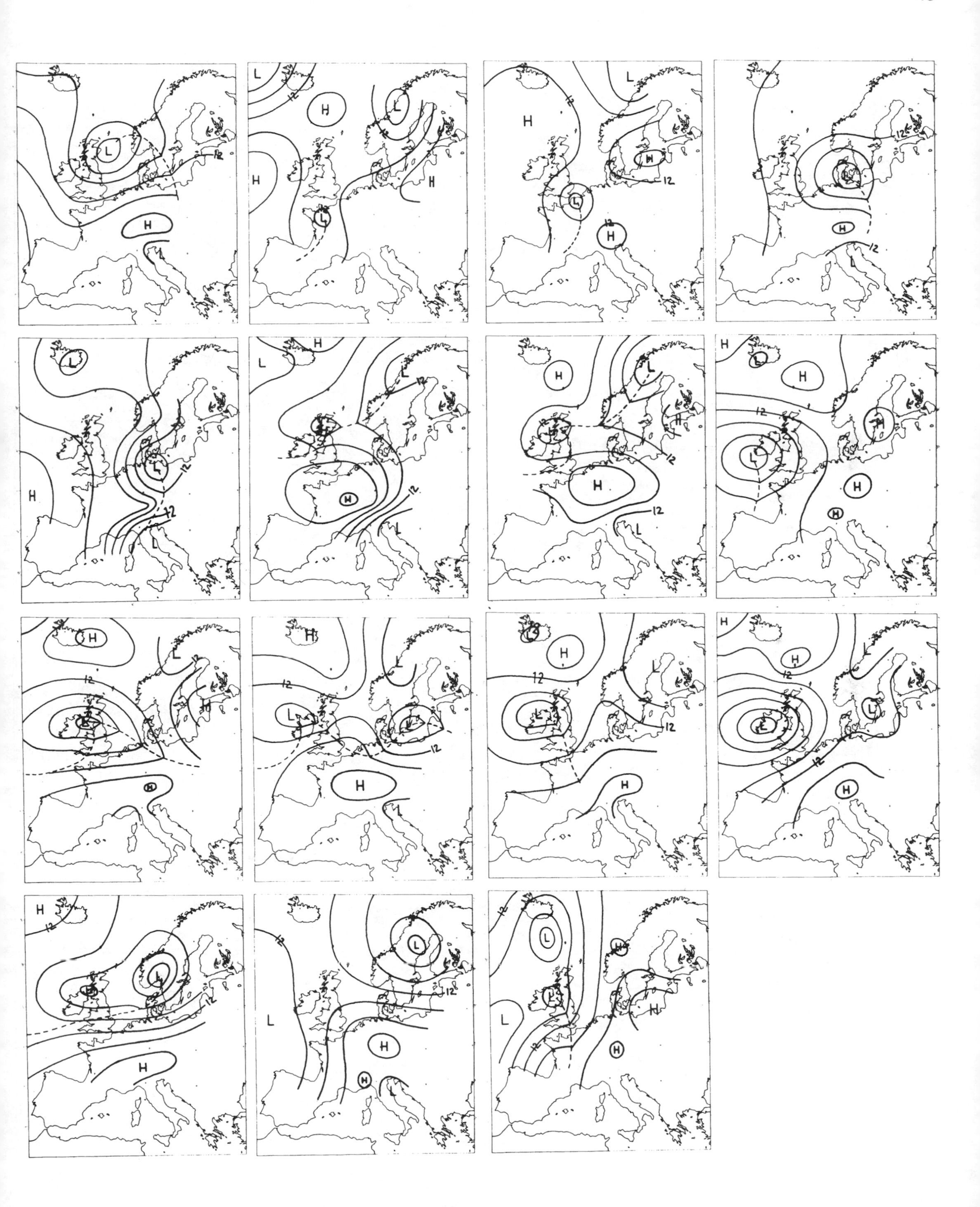

1781 SEPTEMBER

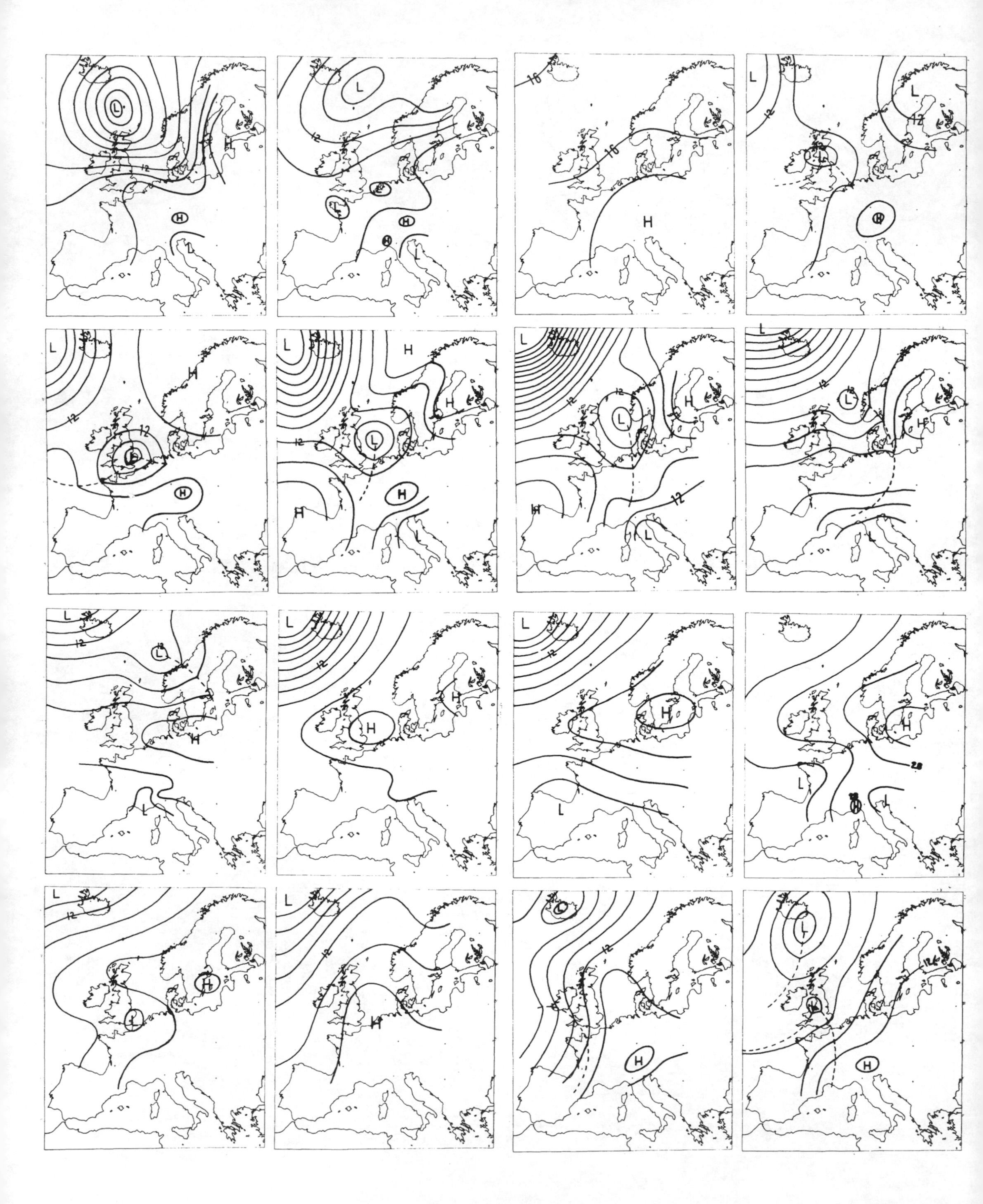

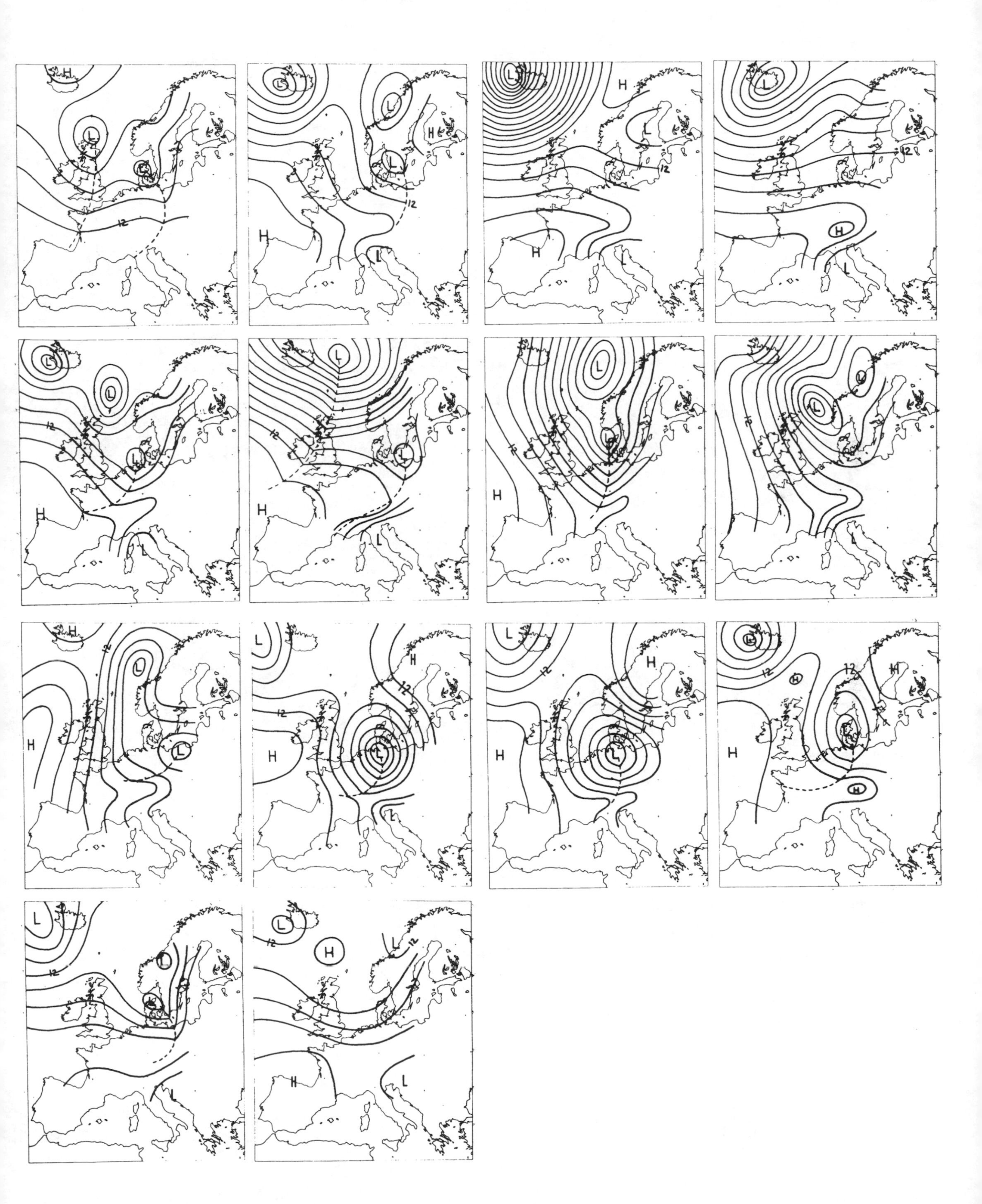

1781 OCTOBER

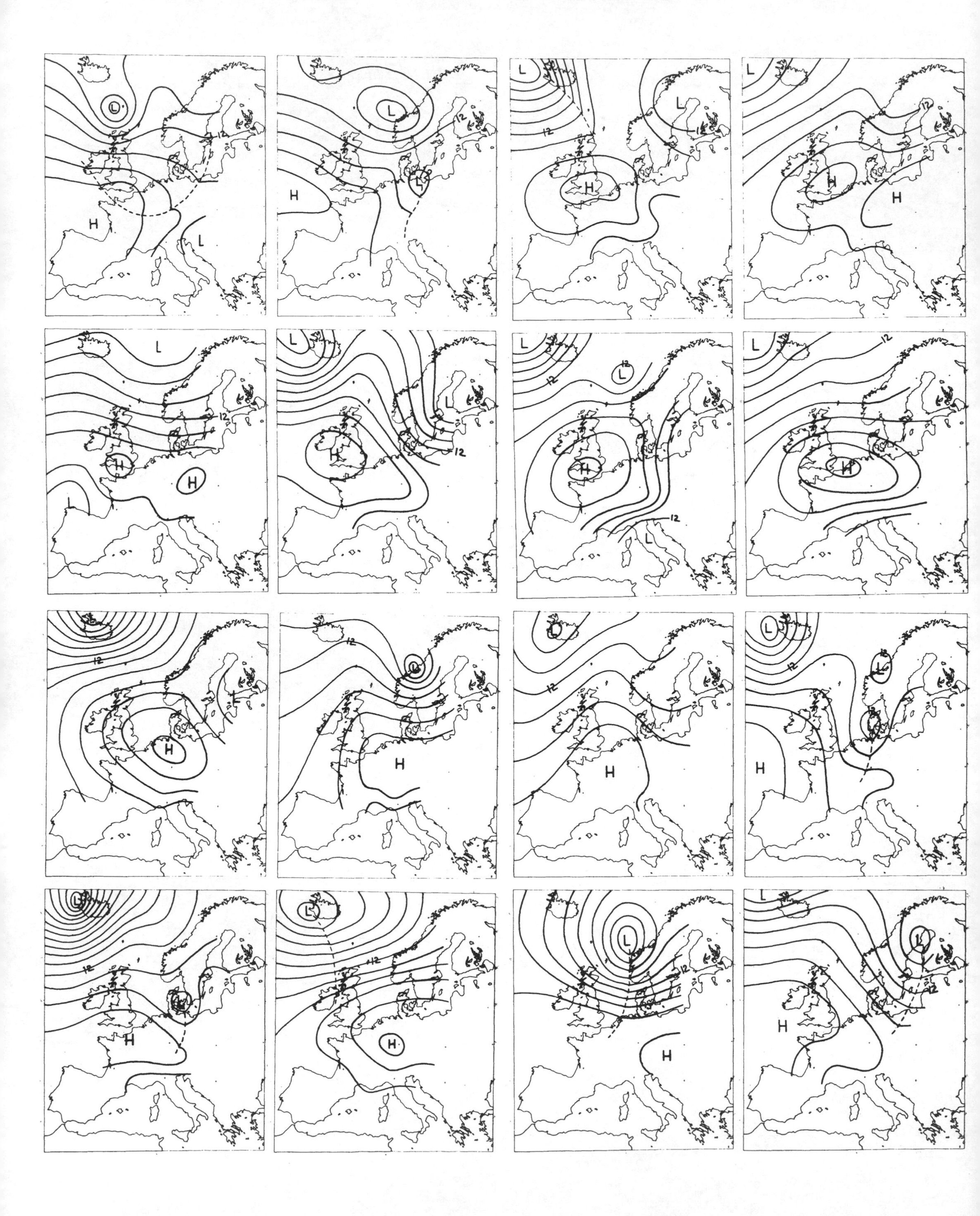

1781 NOVEMBER

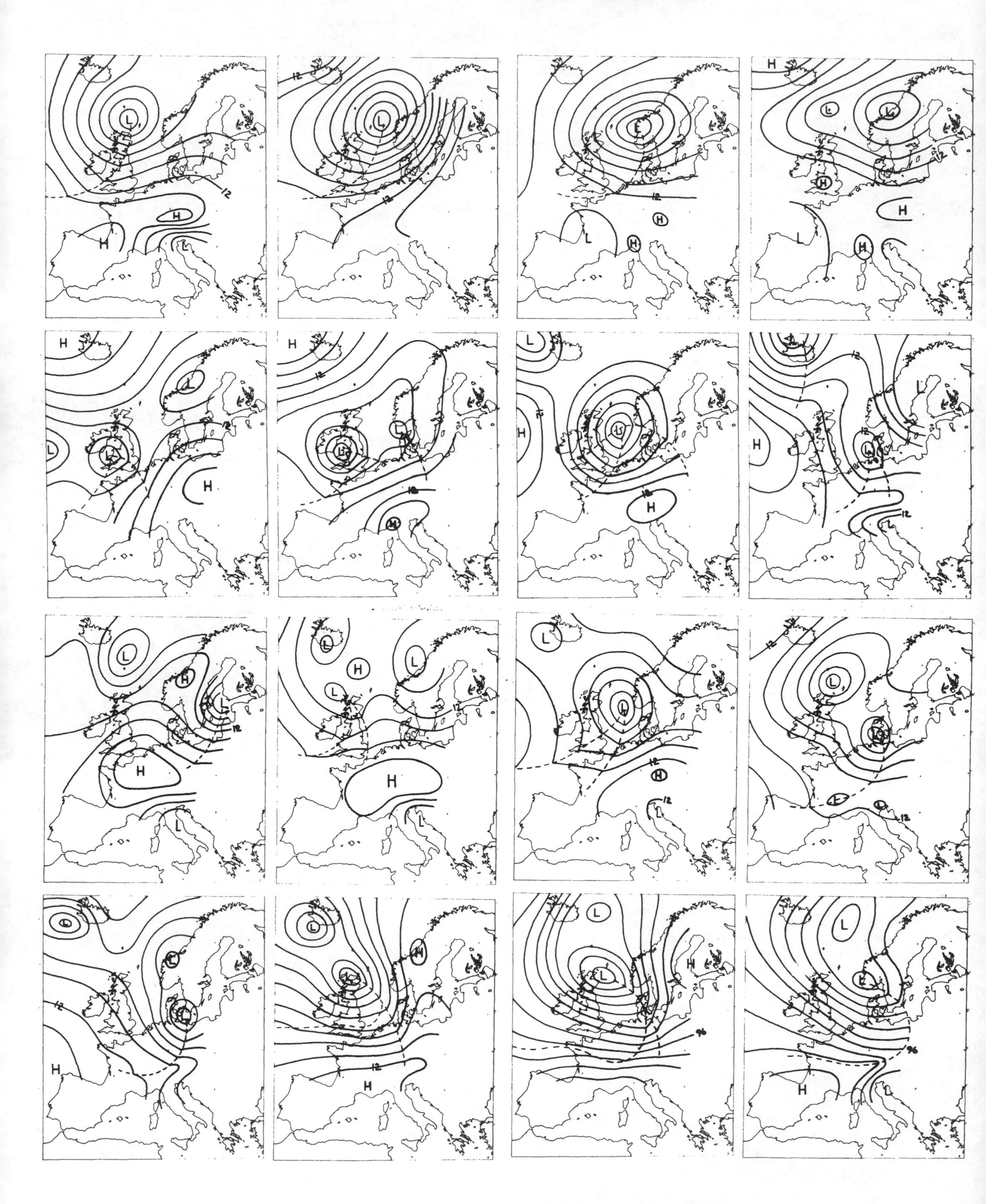

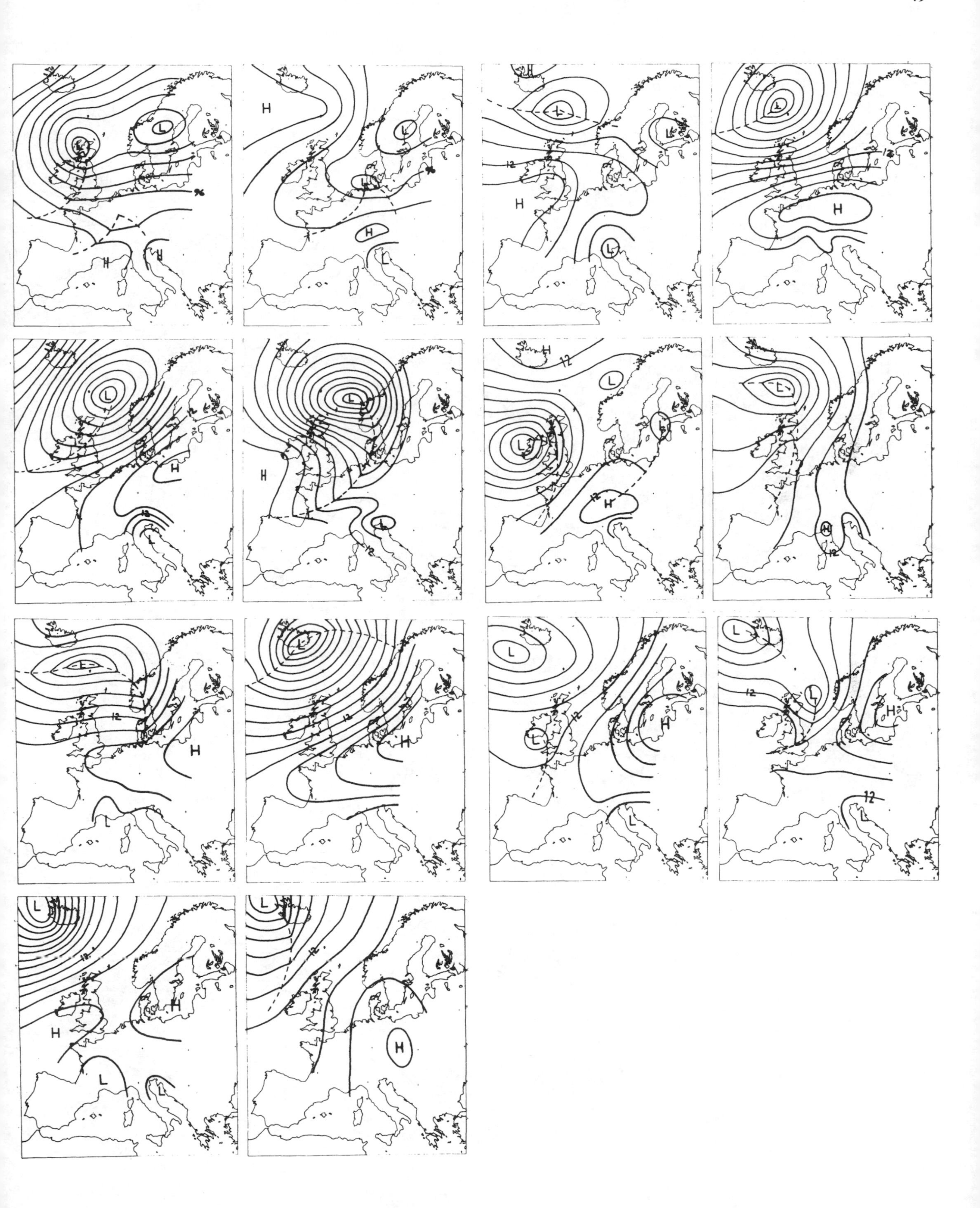

1781 DECEMBER

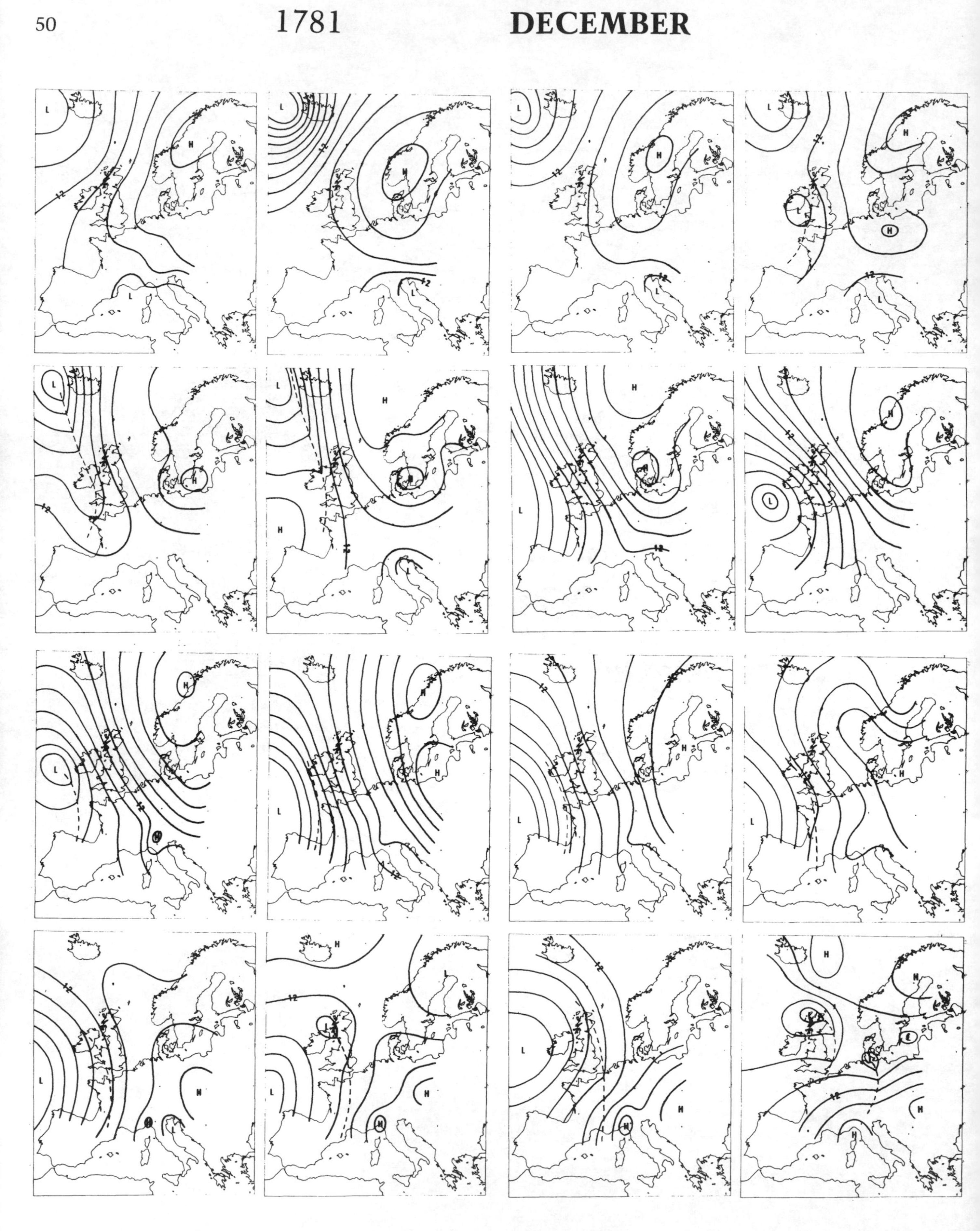

L
H
12
L
H
12
H
L
H
12
H
H
L
H
12
H
H
L
H
12
L
H
12
H
L
12
H
L
12
H
L
12
H
L
12
H
L
L
12
H
H
L
L
12
H
H
L
12
H
L
L
12
H
L
H
L
12
H

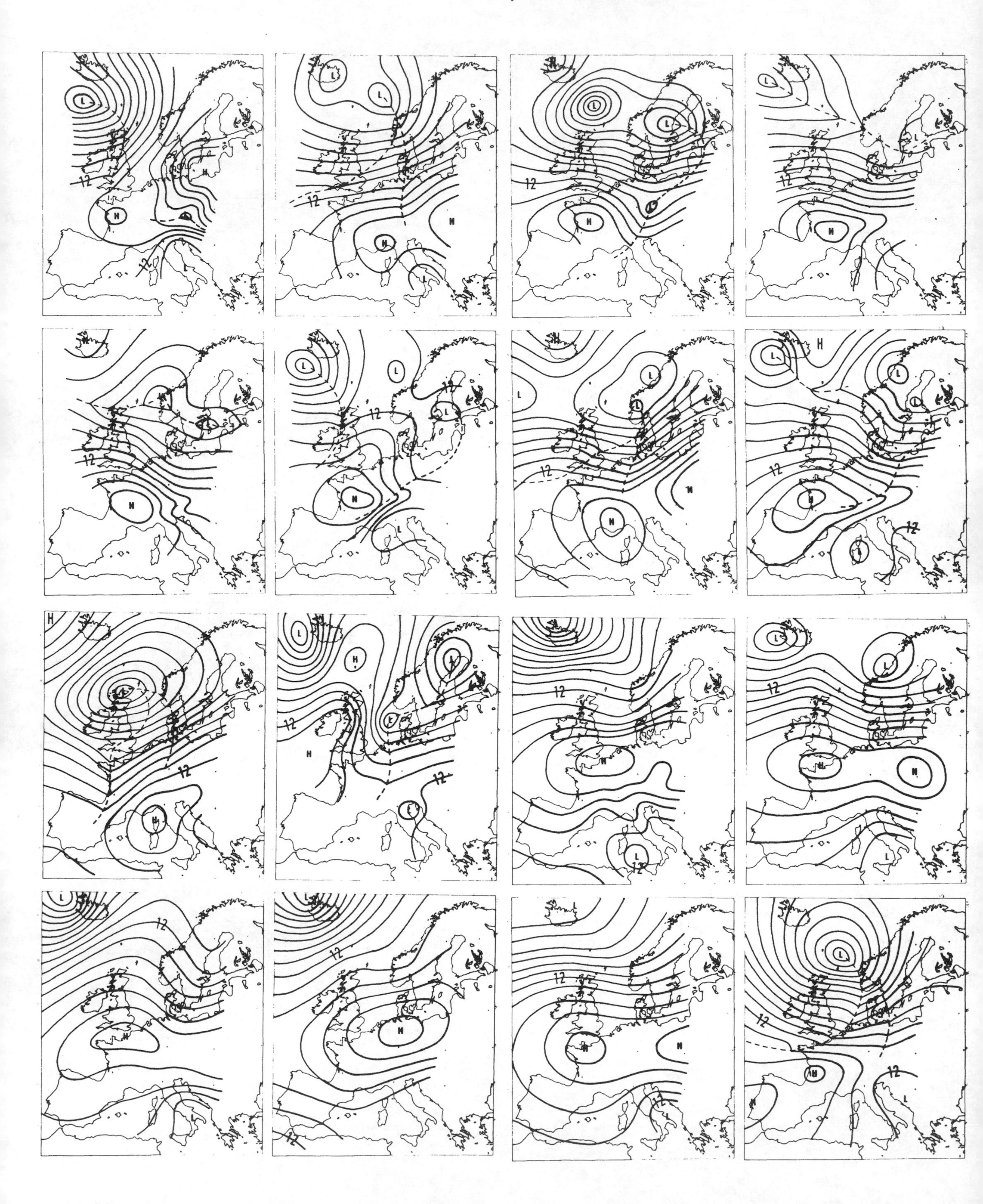
12
L
H

12
96
H
L

1782 FEBRUARY

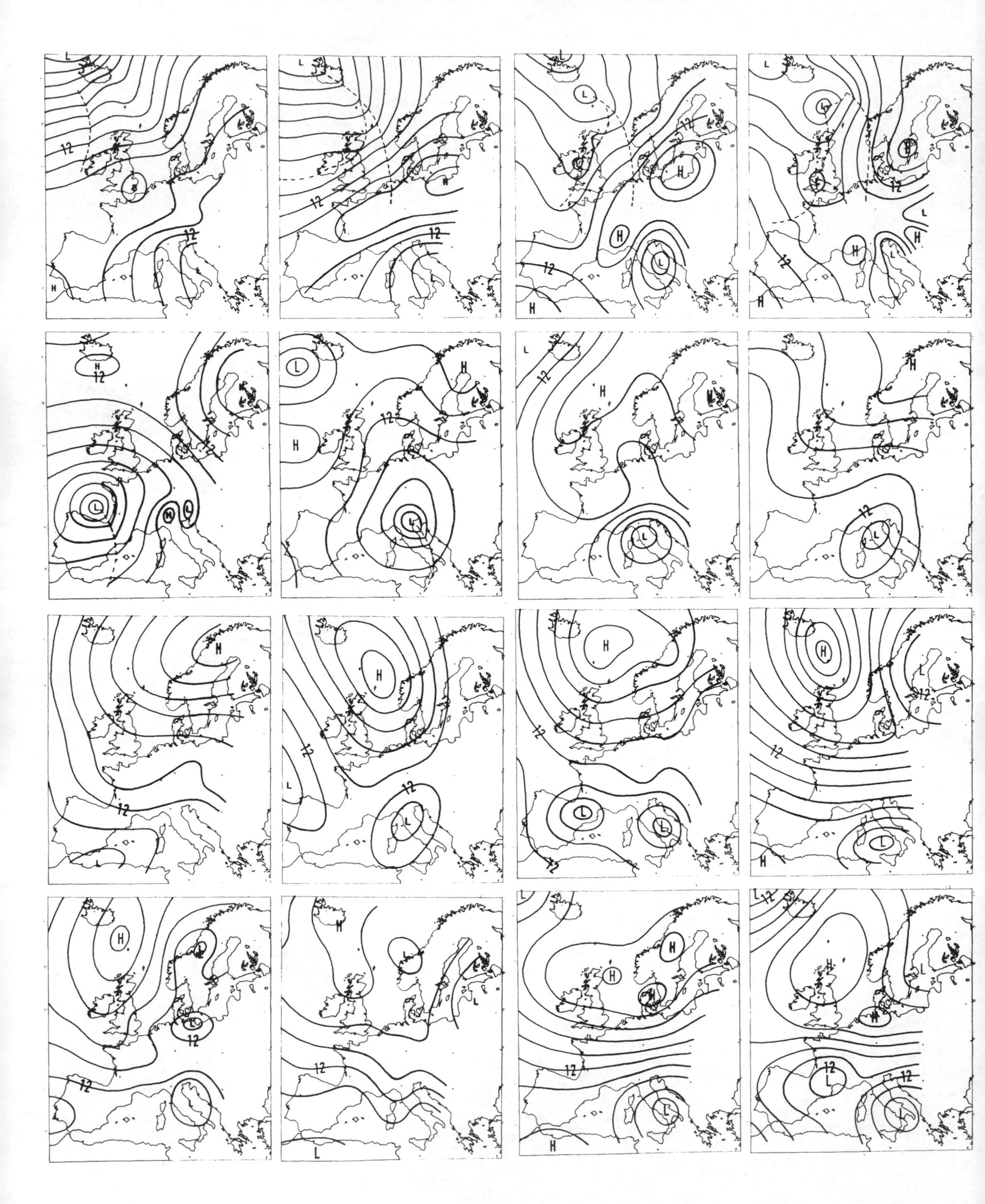

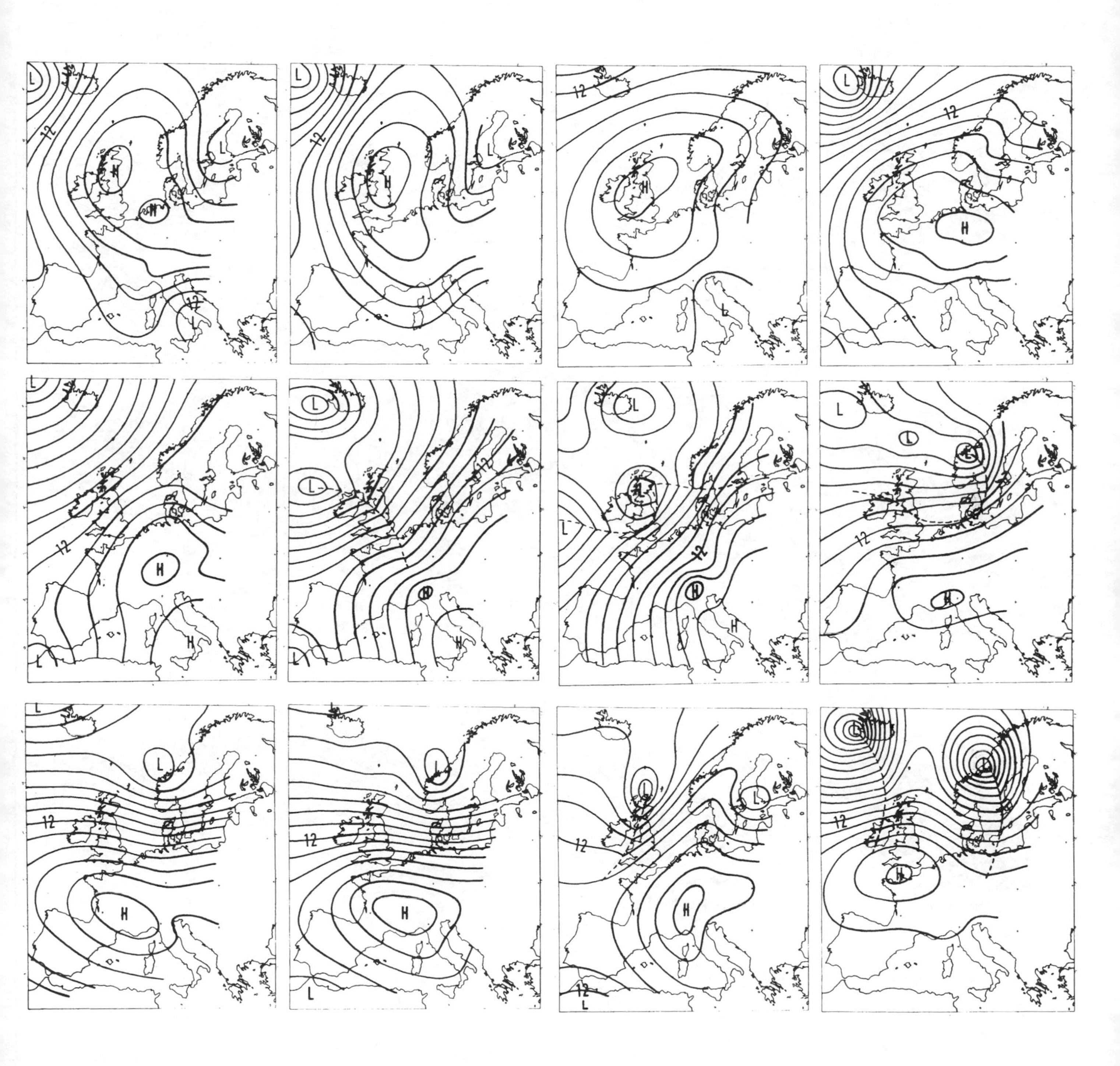

L
12
H
H
L
12
L
L
12
H
L
H
L
12
L
12
H
L
12
12
H
H
L
L
L-
12
H
H
L
L
L
L
12
H
H
L
L
L
12
H
L
L
12
H
L
L
12
H
L
L
L
12
H
12
L
L
L
12
H

1782 MARCH

1782 APRIL

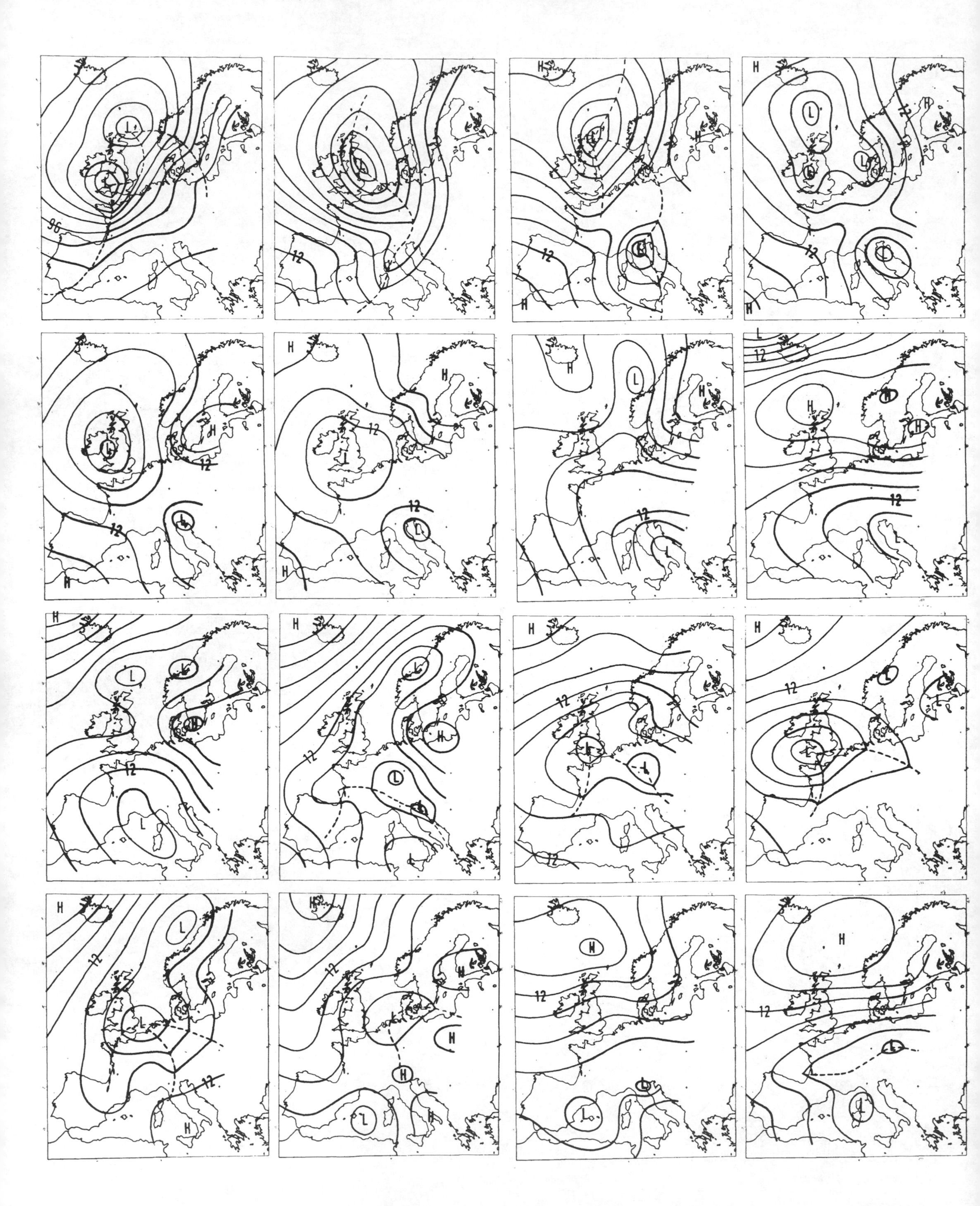

L
H
12
L
H
12
L
12
H
12
L
H
12
L
12
H
12
L
H
12
H
L
12
L
H
L
12
H
H
12
H
L
12
L
H
12
H
H
12
L
L
H
12
H
12
L
12
H
12
L
L
12
L
H
12
L
12
H
L

1782 MAY

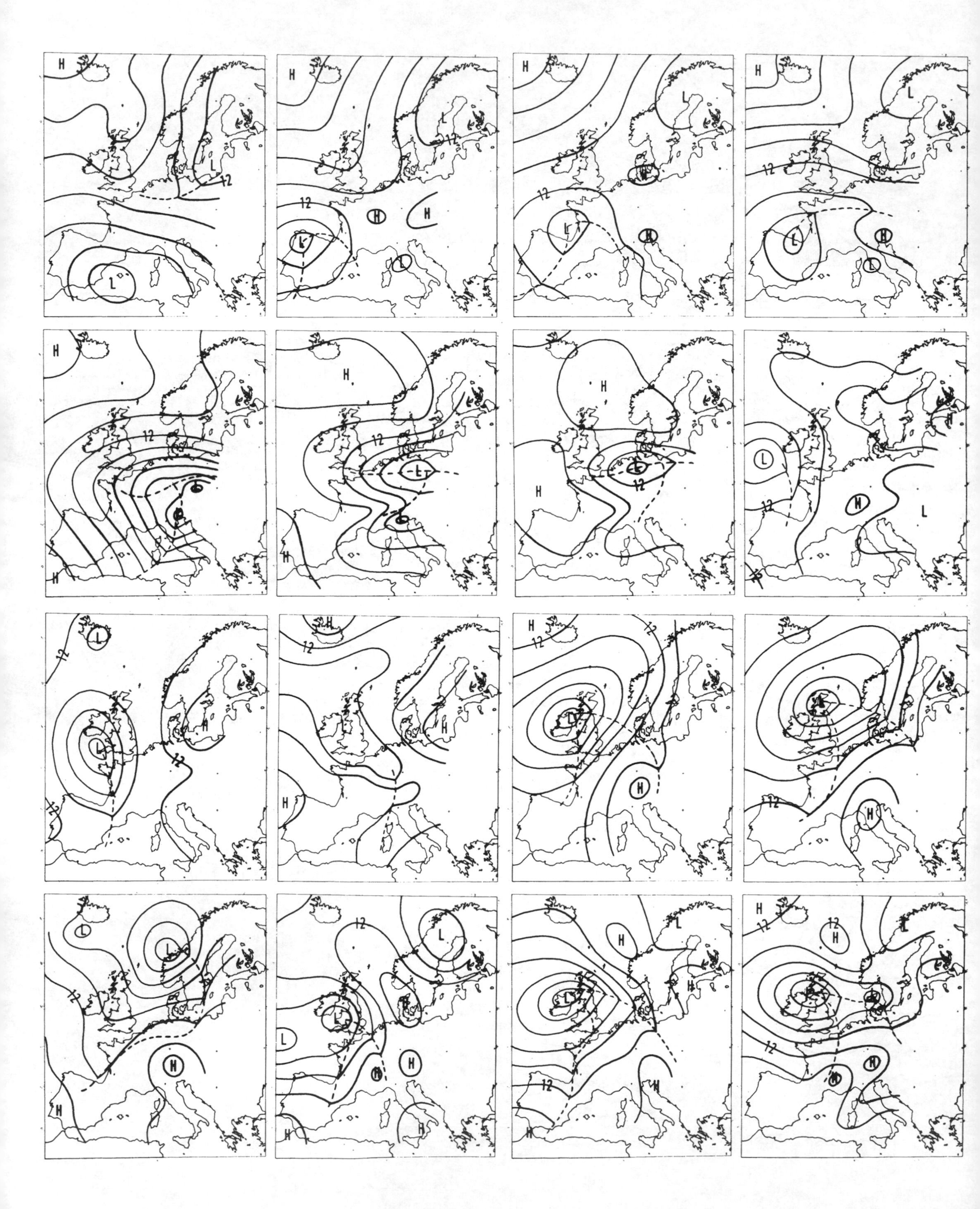

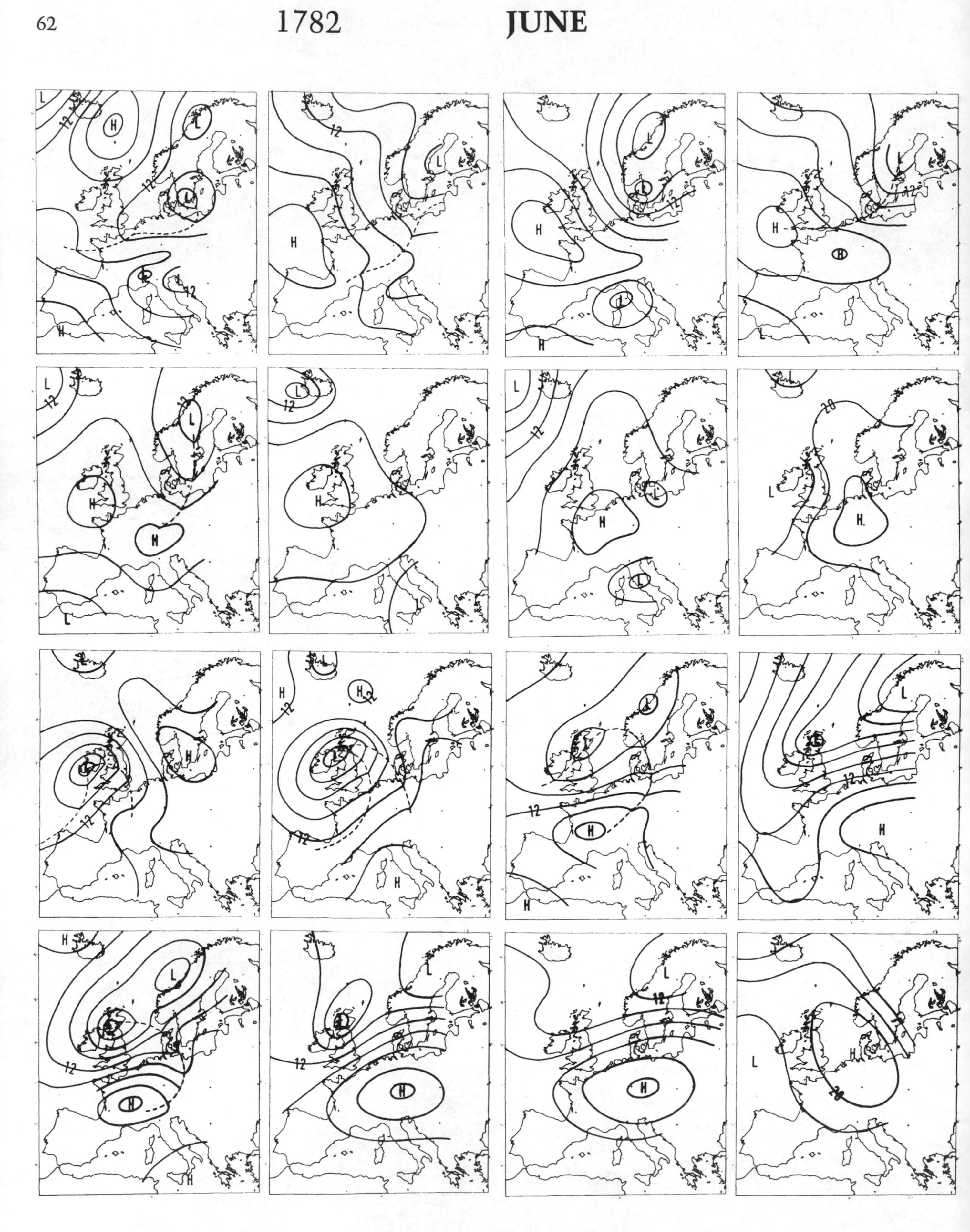

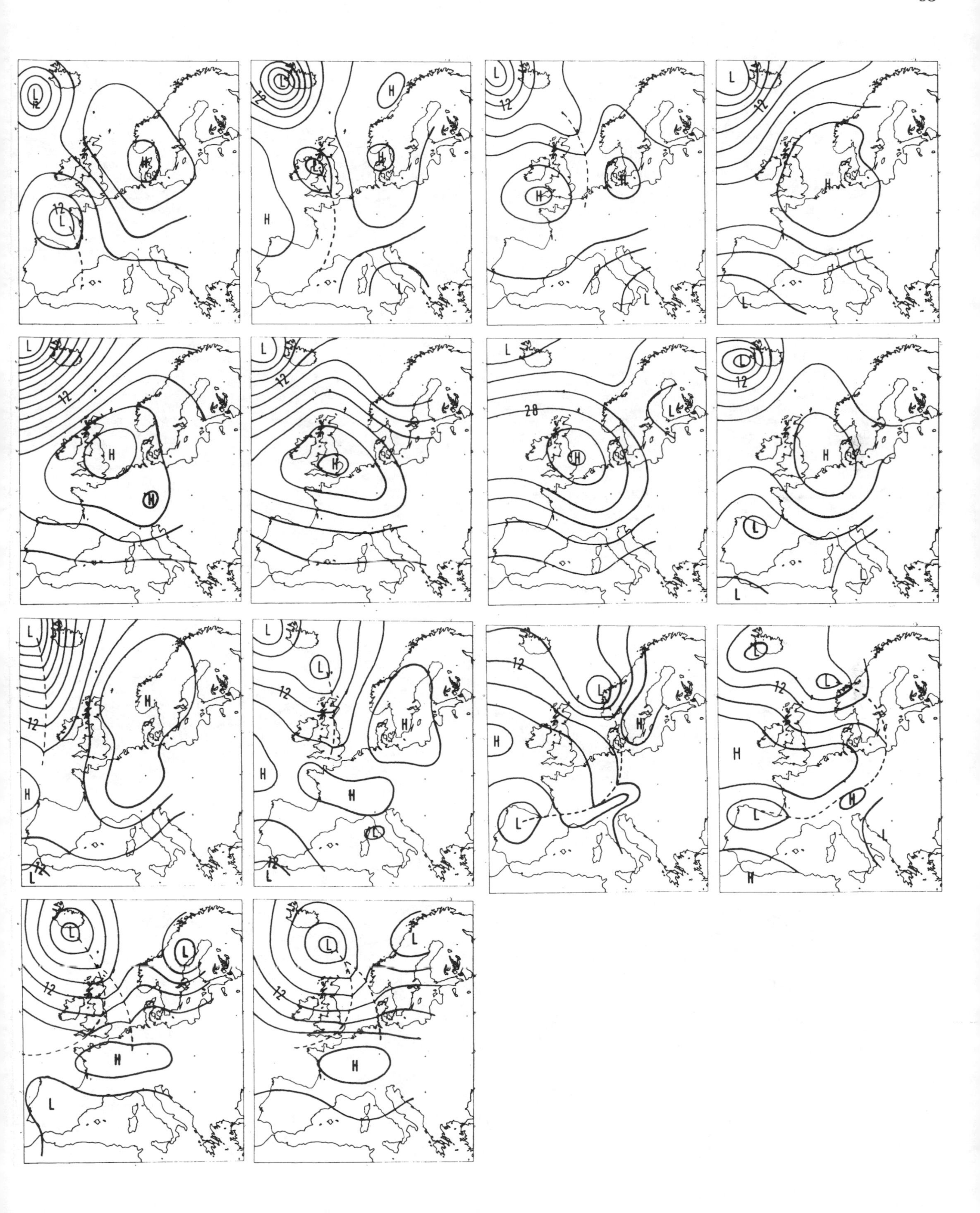
L
12
H
12
L
L
12
H
H
L
H
L
L
12
H
H
L
L
12
H
L
L
12
H
H
L
12
H
L
28
L
H
L
12
H
L
L
L
12
H
H
L
12
L
L
12
H
H
H
L
L
12
12
L
H
H
L
L
12
L
H
H
L
L
H
L
12
L
H
L
L
L
12
H

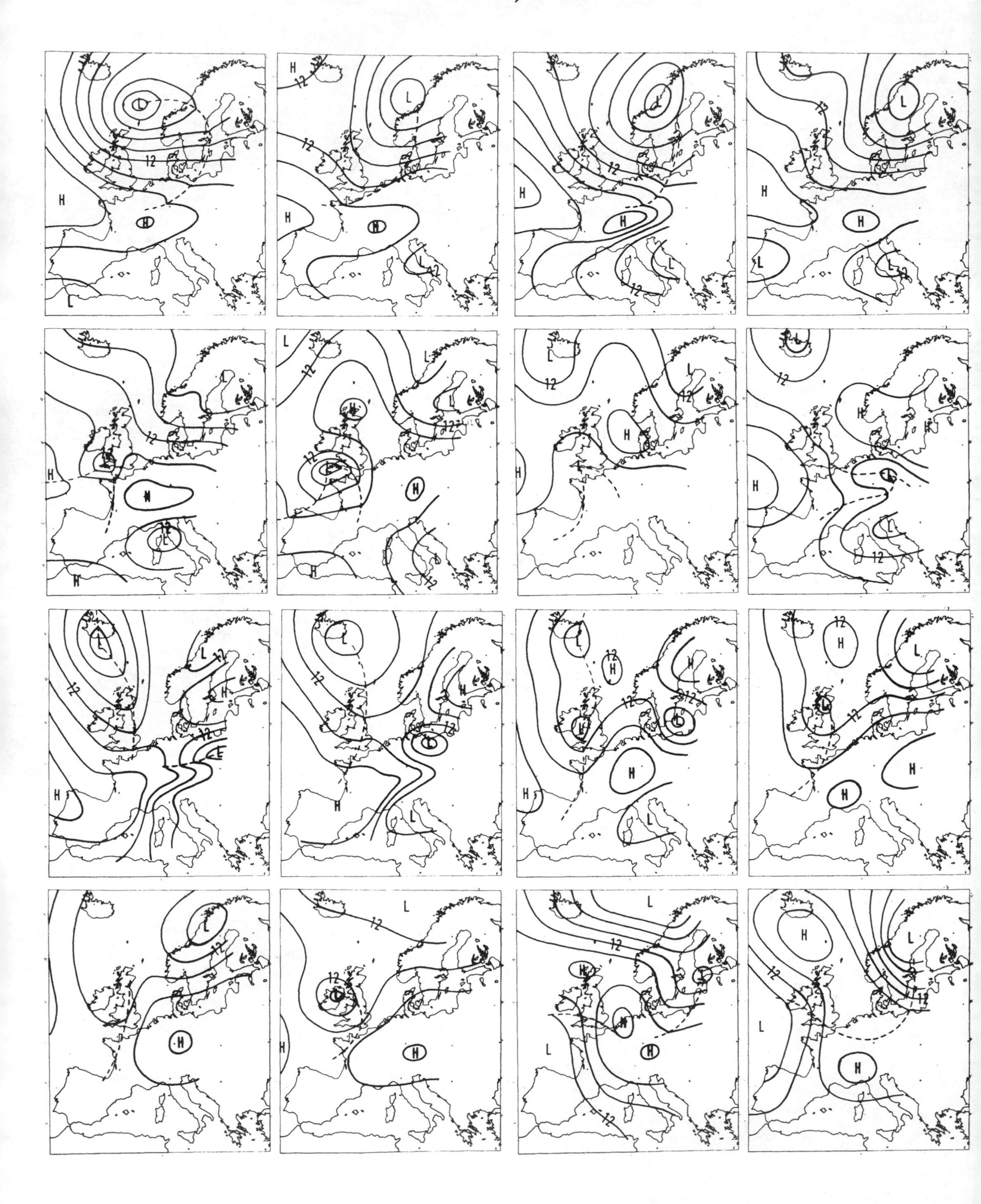

H
L
12

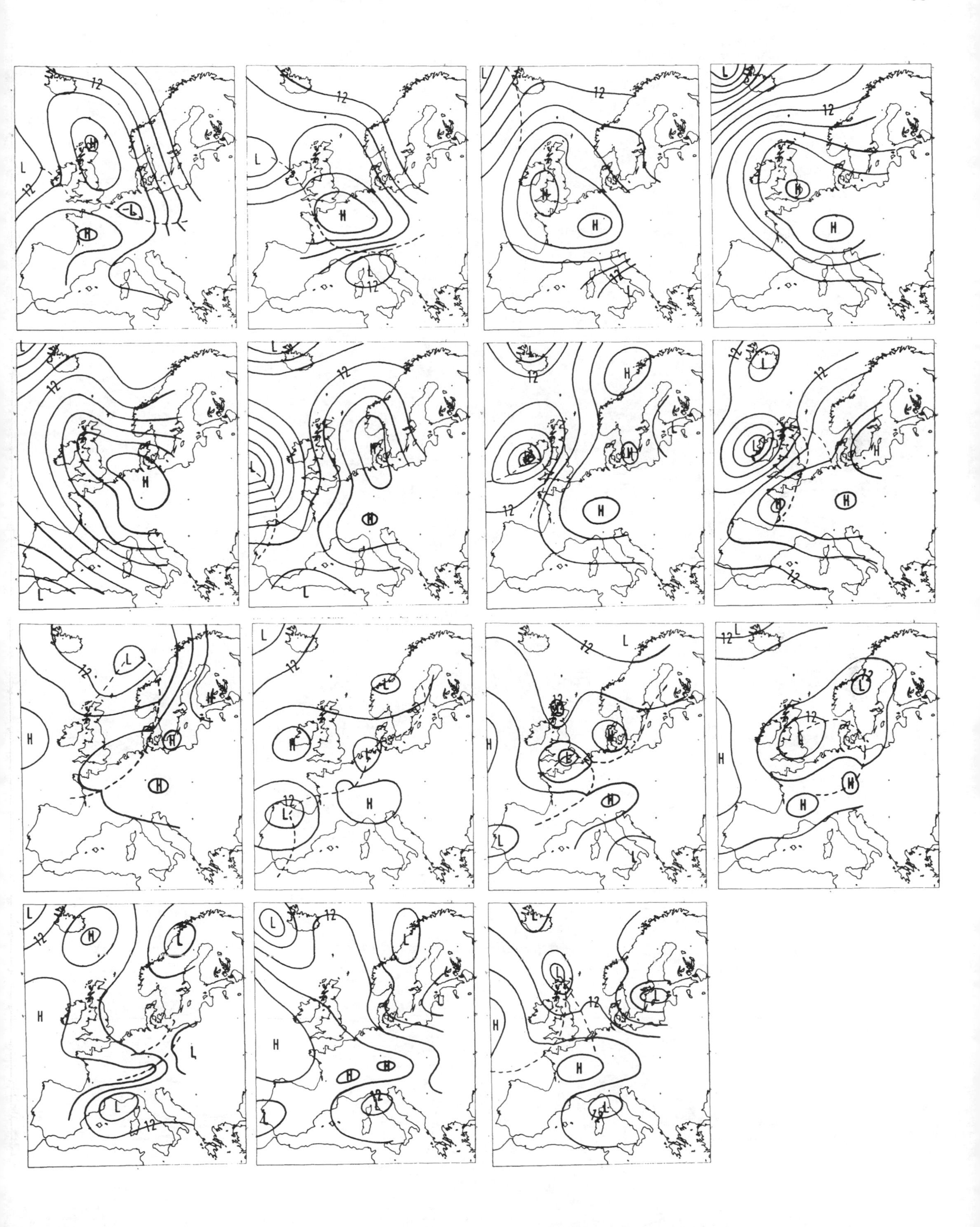

1782 AUGUST

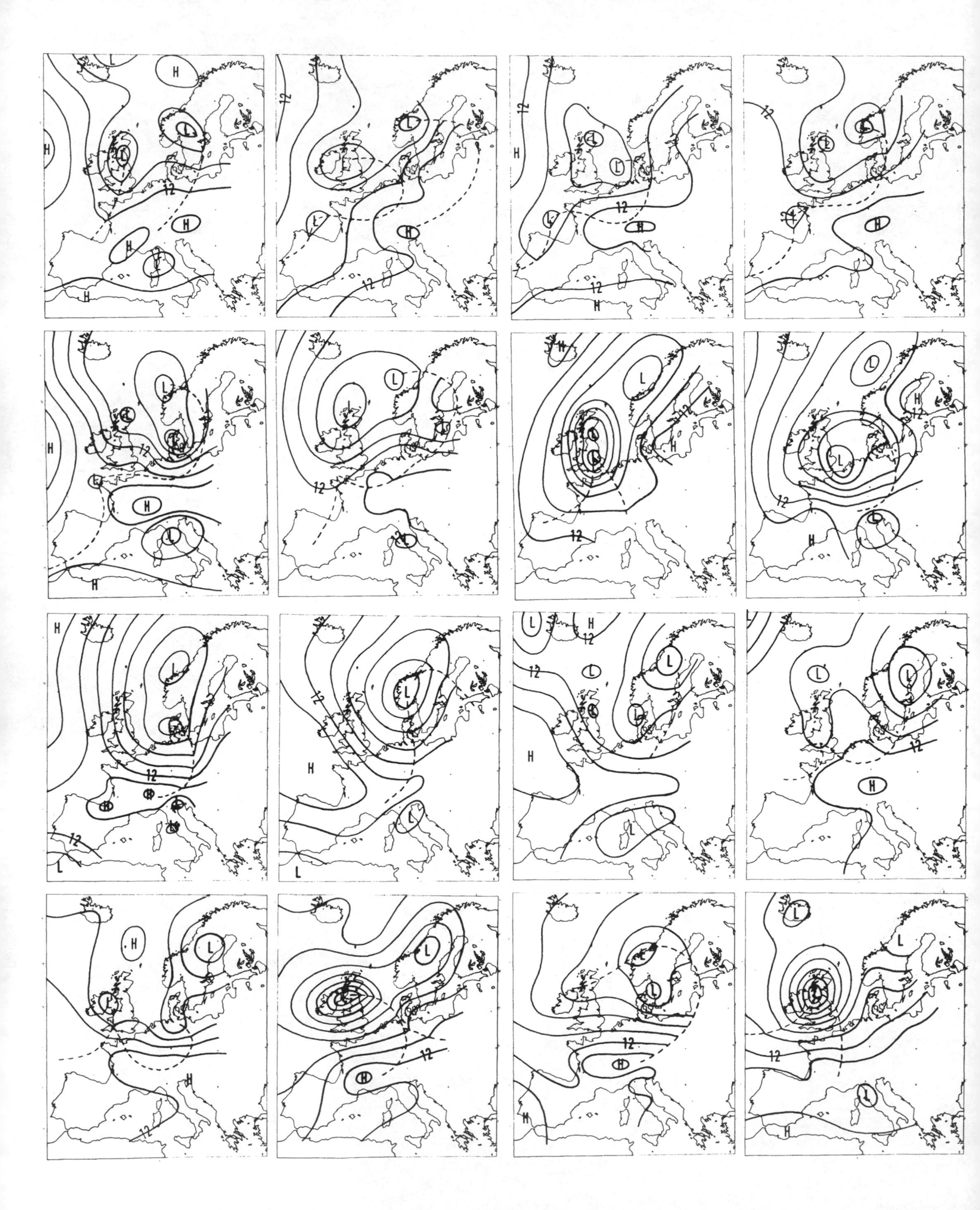

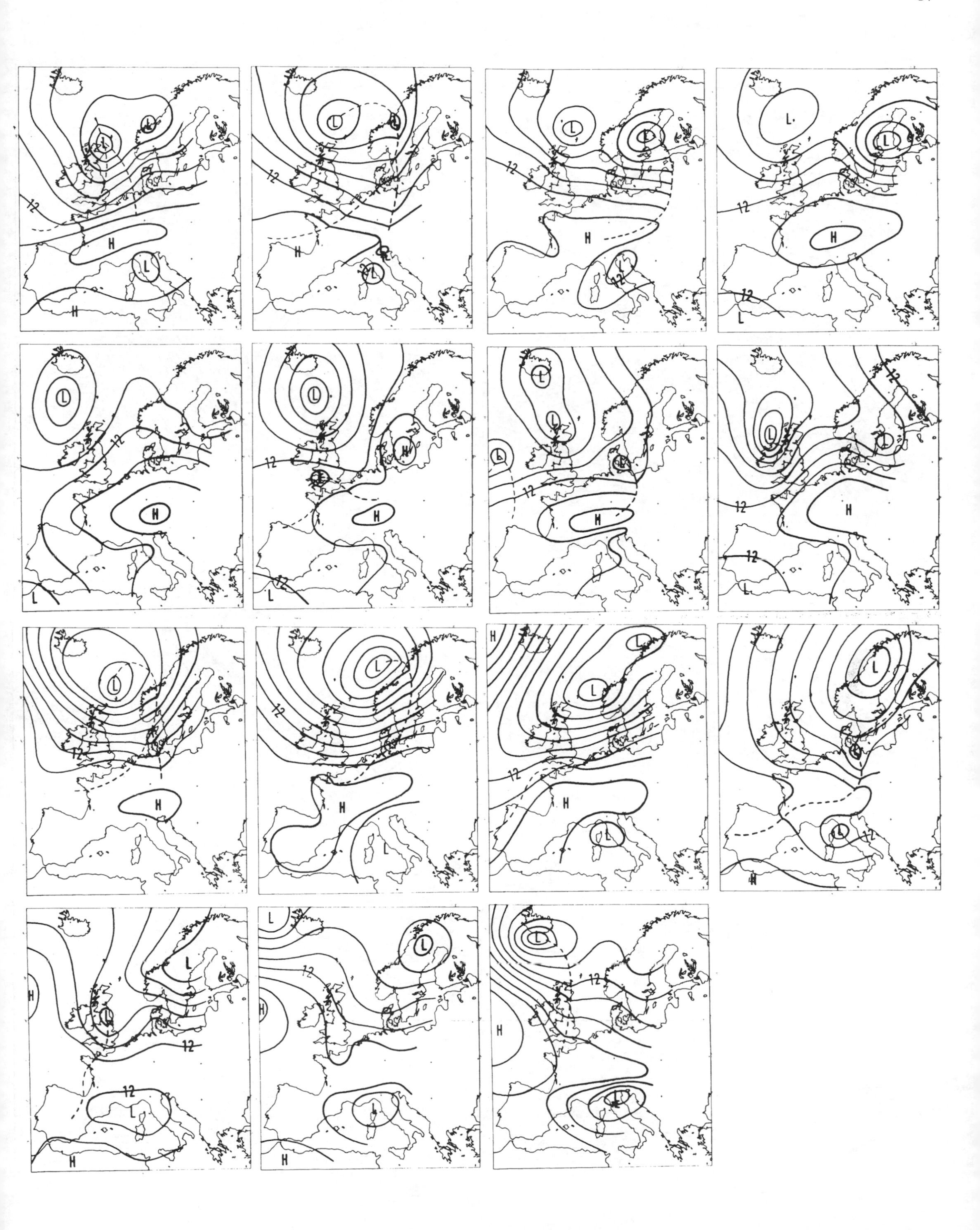

1782 SEPTEMBER

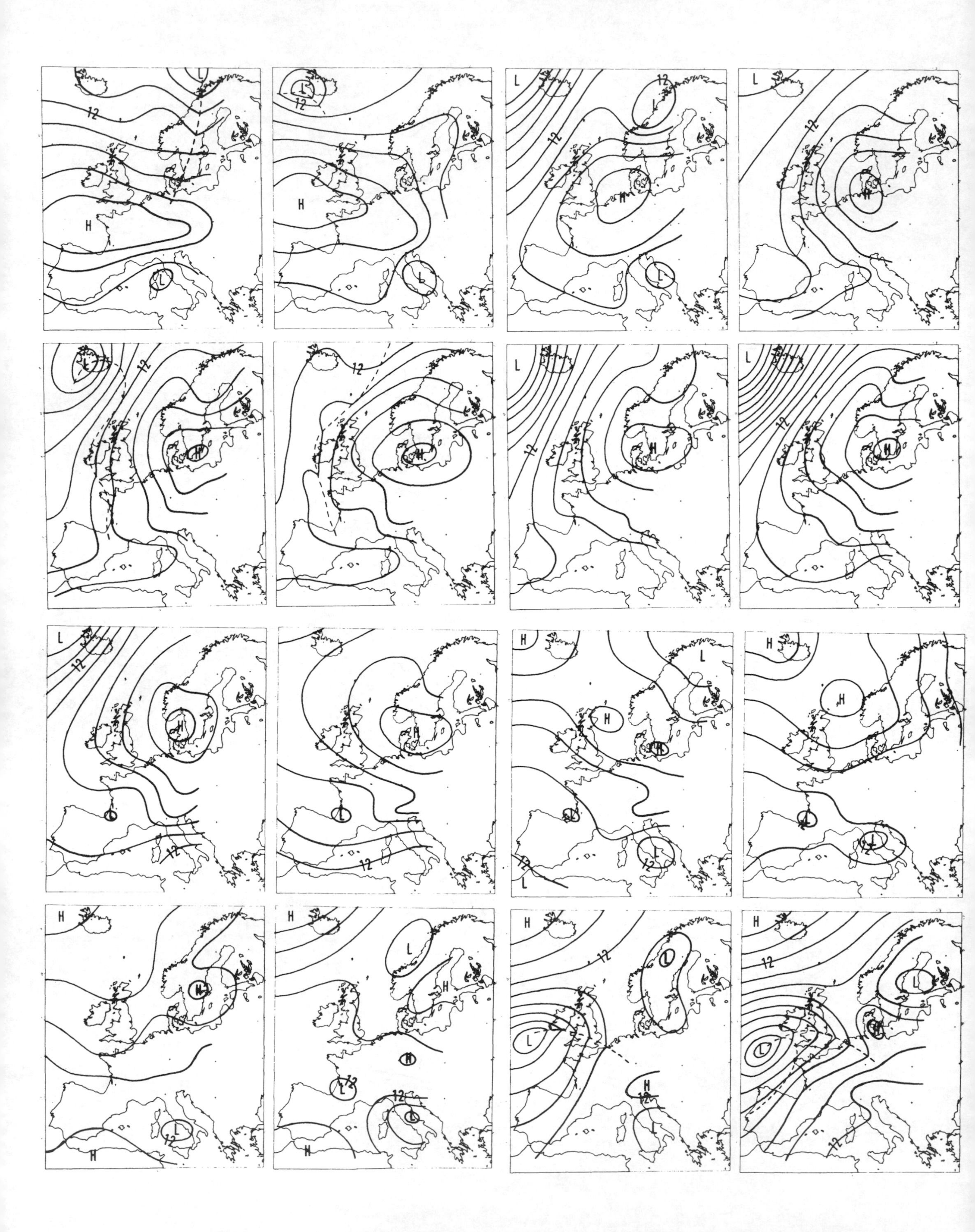

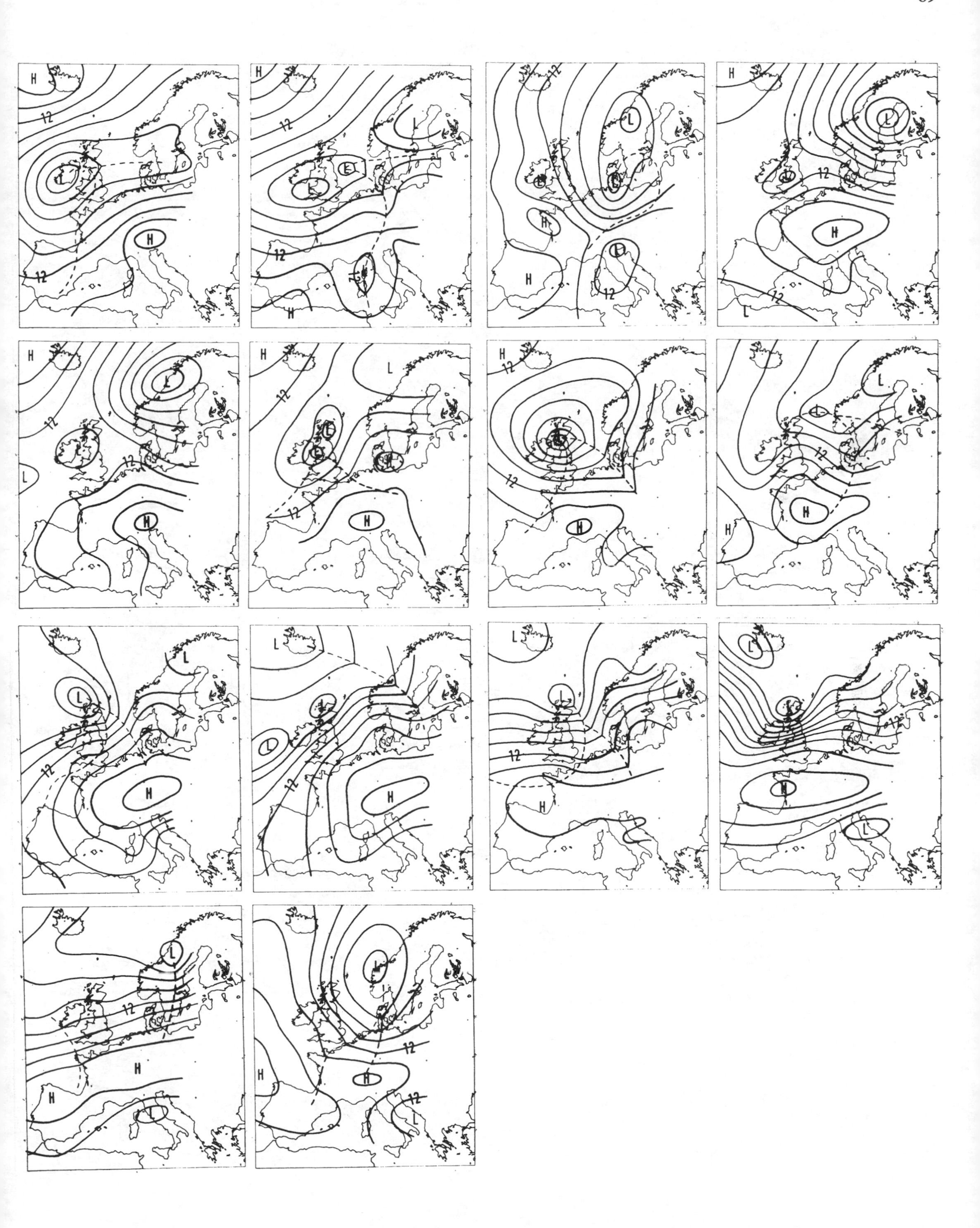
H
12
L
H
12
L
H
12
H
L
12
H
L
12
H
H
L
12
H
L
H
12
H
L
12
H
L
H
12
H
L
12
H
L
12
H
L
H
12
H
L
12
H
L
L
12
H
L
12
L
H
H
12
L

L
12
H
L
12
H
L
12
H
H
L
12
12
L
12
H
12
L
H
12
L
H
L
12
H
L
12
L
12
H
L
12
12
H
L
12
L
12
H
L
12
H
12
12
H
L
12
12
L
H
12
12
H

H
12
L
H
L
12
H
12
12
H
12
H
H
12
12
L
H
12
L
H
L
12
H
H
12
H
L
L
12
12
H
L
12
L
H
L
12
H
L
L
12
12
12
H
L
12
L
L
12
12
H
L
12
H
L
H
12
H
H
12
L
L
L
12
H
L
L
12
H
12
H
12
H
L
L
12
H
L
L
12
12
L

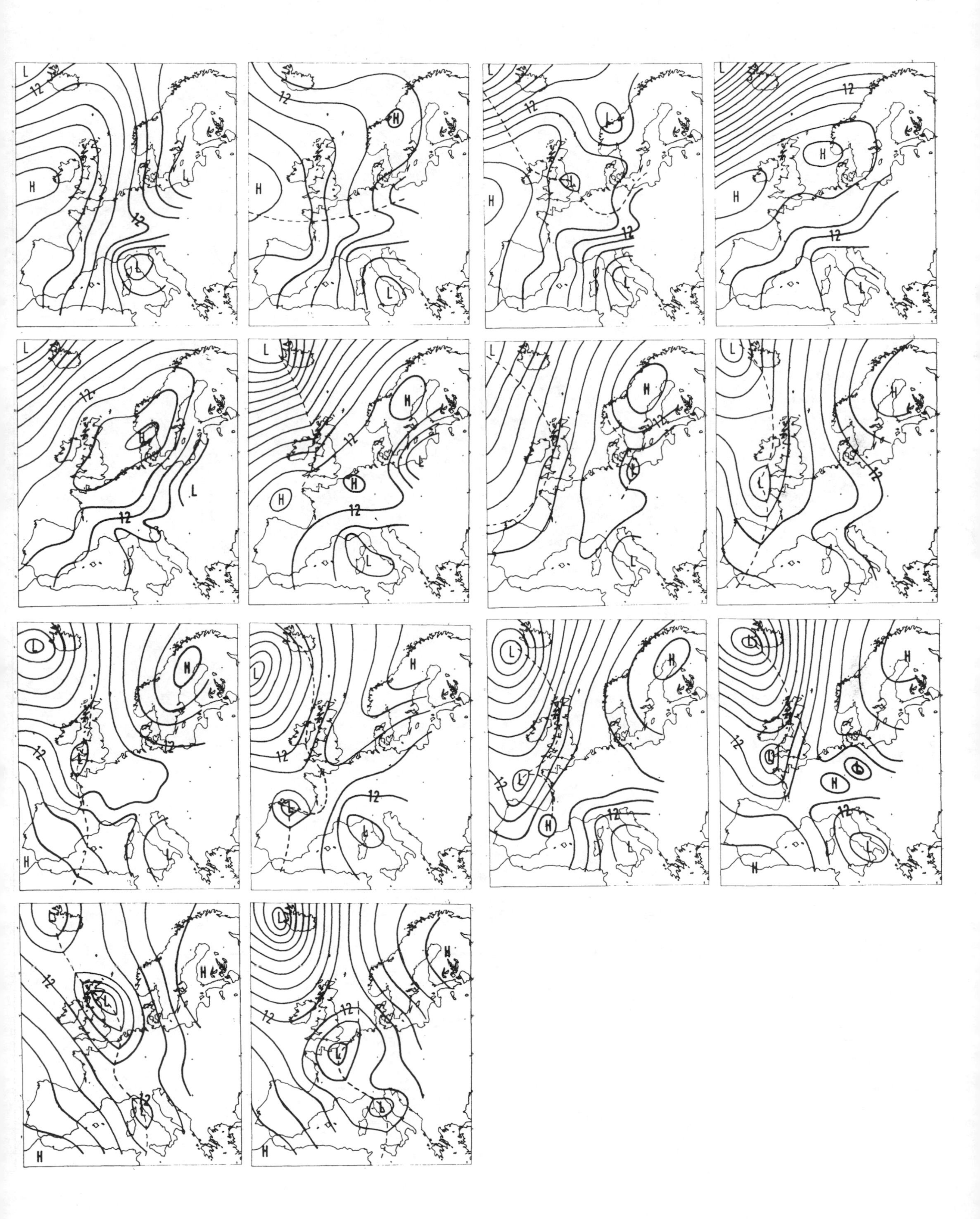

L
12
H
L
12
L
12
H
H
L
L
12
L
H
L
12
L
L
12
H
H
12
L
L
12
L
12
H
L
12
H
H
L
L
H
12
L
L
H
L
12
L
12
H
H
12
L
12
H
L
12
L
H
H
L
L
12
L
12
H
L
12
H
L
H
12
L
H
L
12
L
H
L
12
L
H
L
12
H

12
H
L

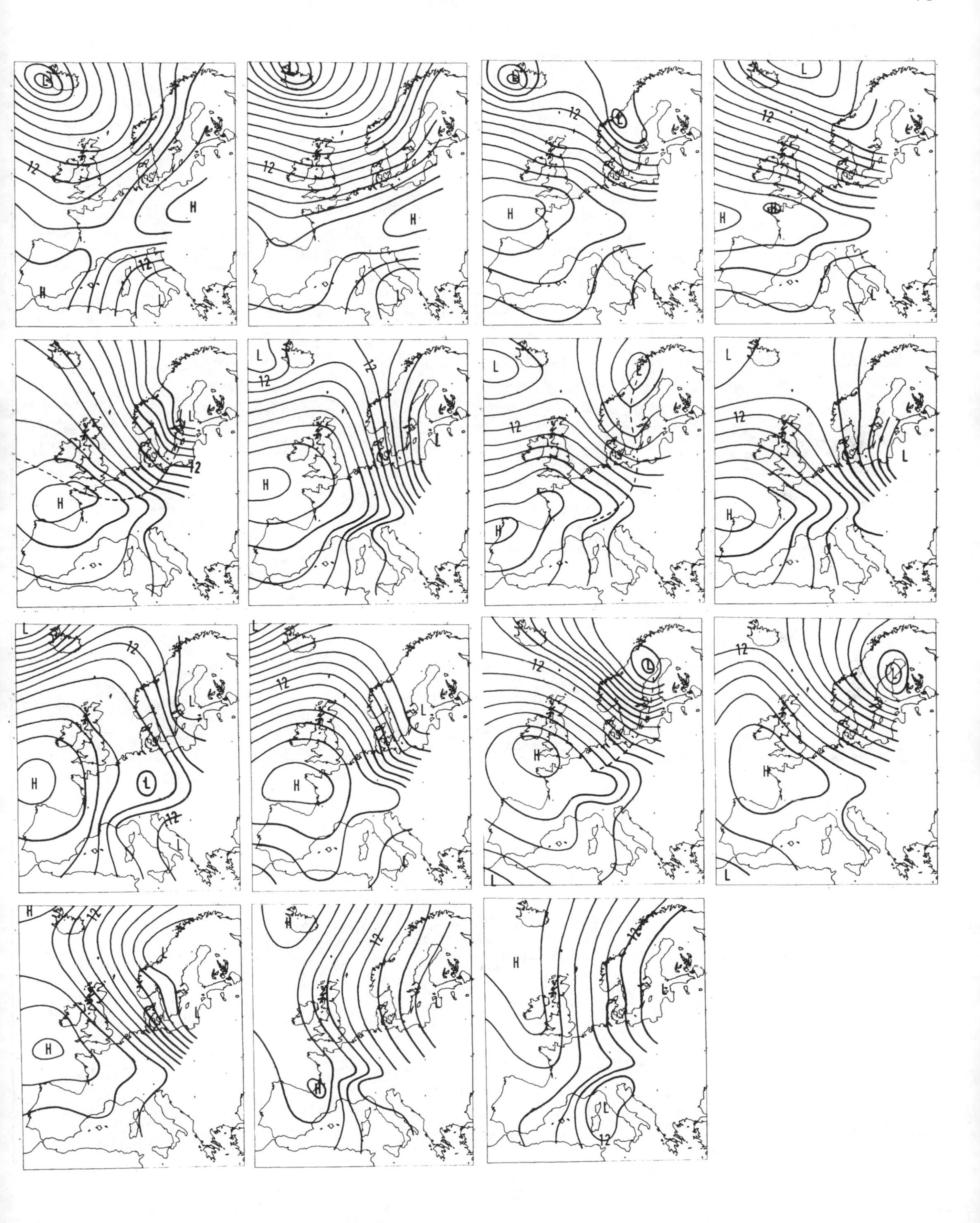

12
H
H
L
12
H
12
H
12
H
12
H
L
L
12
H
H
L
12
L
12
H
L
12
H
L
L
12
H
L
12
L
L
12
H
L
12
L
H
L
12
L
H
L
L
H
12
H
L
H
12
H
H
12
L
12

H
L
12
H
H
L
12
L
12
H
L
L
12
H
H
L
12
H
L
12
H
L
H
12
L
H
L
12
L
L
12
H
H
L
L
L
12
H
L
L
L
12
H
H
L
L
L
12
H
H
L
12
H
L
12
H
H
L
12
H
H
12
L

H
12
L
H
12
L
12
L
H
12
H
L
H
12
H
12
H
L
L
H
12
L
12
H
L
L
H
12
L
H
L
12
H
L
H
12
H
H
L
12
L
H
12
L
H
H
12
L
H
L
H
L
12
L
H
H
12
L
H
12
H

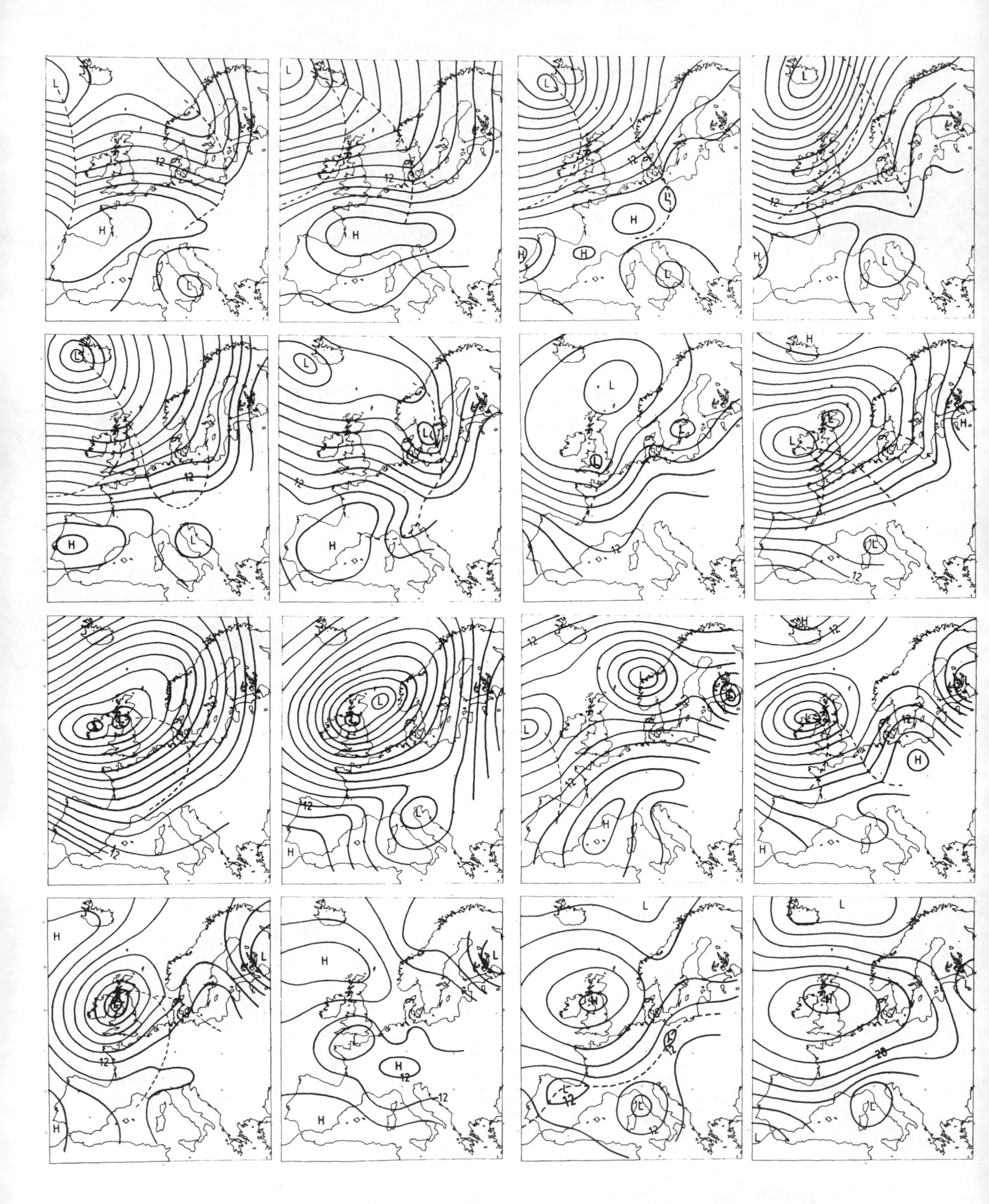

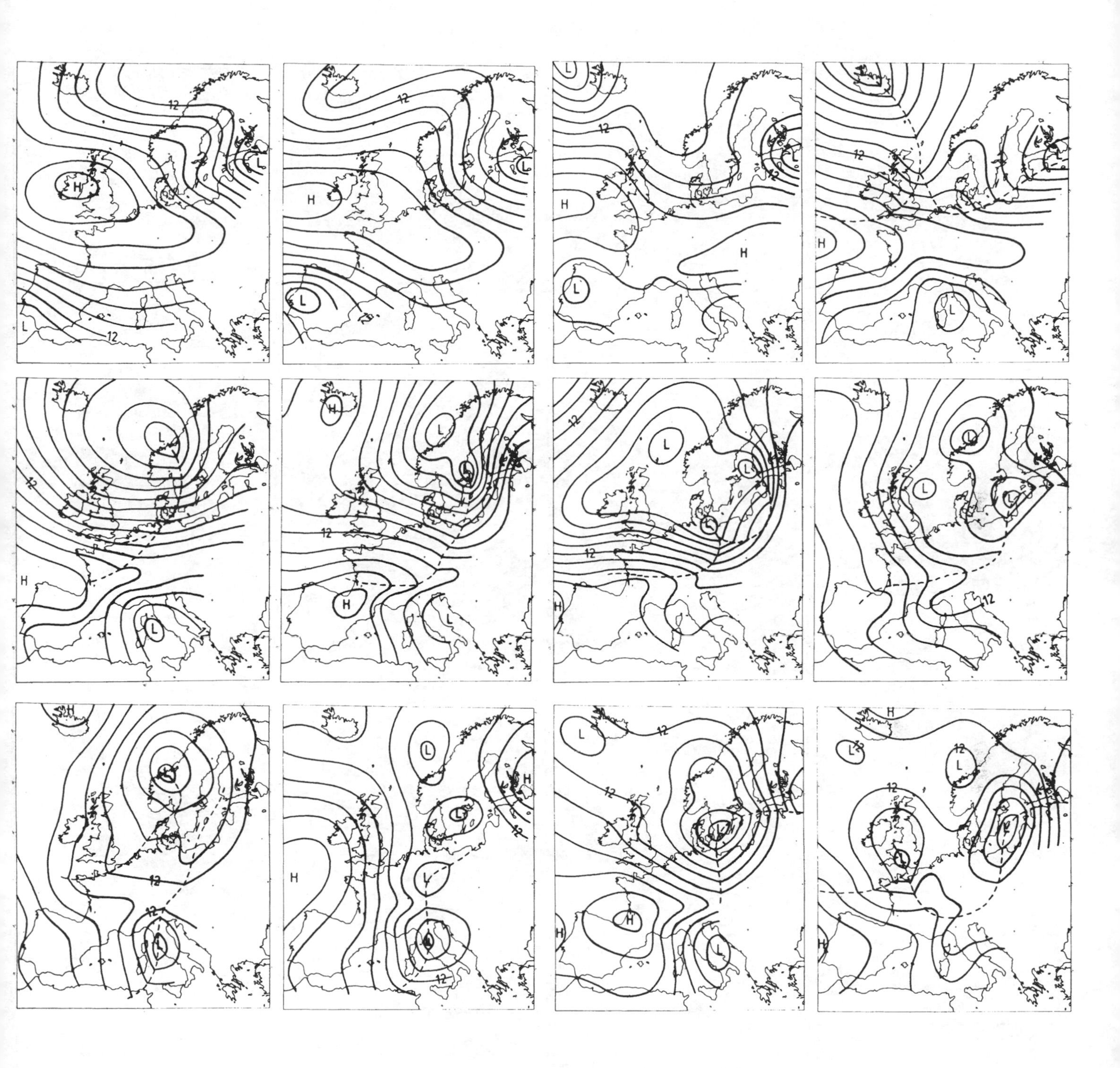
H
L
12
L
H
12
L
L
12
H
L
H
L
12
H
L
L
12
H
L
H
L
L
12
H
L
12
L
L
L
12
H
L
L
L
L
12
H
L
12
L
12
H
L
L
L
12
H
L
L
12
H
H
L
H
L
12
H
L
12
L
H
L
12
L
12
H
L

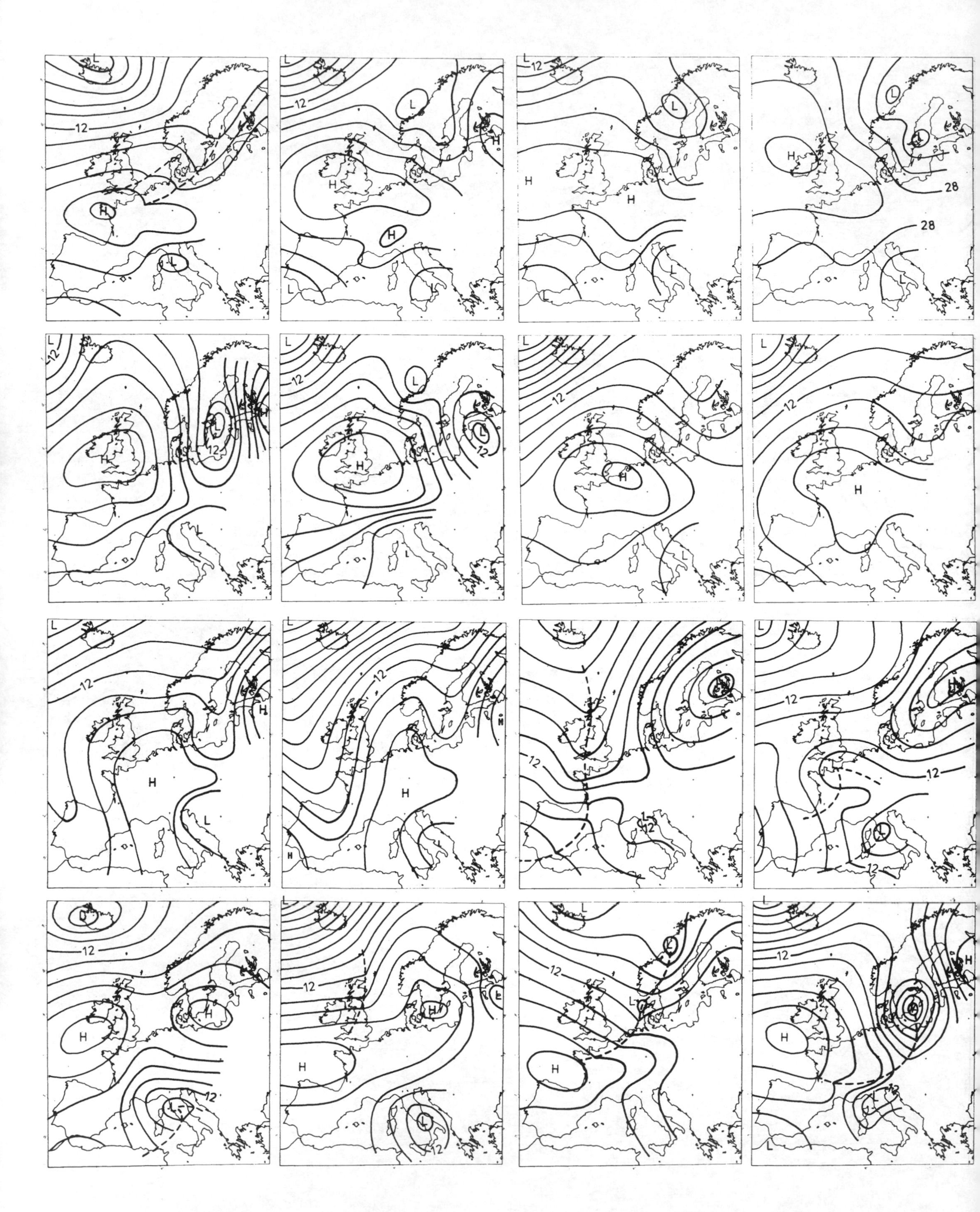
L
12
H
L
12
L
H
H
L
L
12
L
H
H
L
L
L
H
28
28
L
L
12
H
12
L
L
L
12
H
12
L
H
12
L
H
L
12
H
L
12
H
L
L
12
H
H
L
12
L
12
L
12
12
L
L
12
L
12
12
H
L
12
H
L
12
H
L
H
L
12
12
L
L
H
L
12
H
L
12
L
H
L

12
H
L
12
H
L
12
H
L
12
H
L

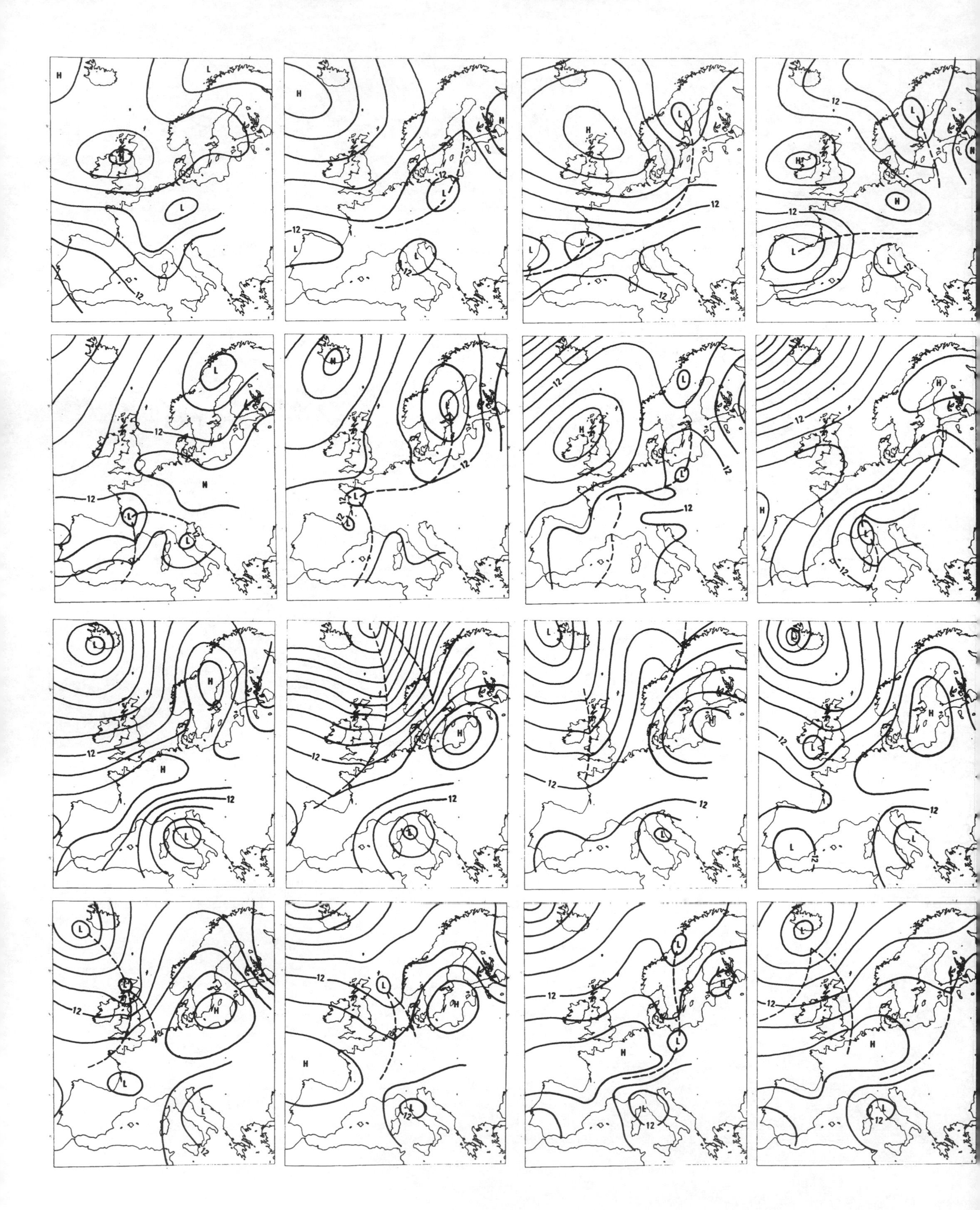
H
L
L
12
12
L
H
12
12
L
L
12
H
L
L
12
L
12
H
L
12
H
L
L
L
L
12
N
L
L
H
12
L
L
12
12
L
L
L
12
H
12
L
12
H
H
L
12
L
H
H
12
L
12
L
H
12
L
12
L
L
12
L
12
L
12
H
L
12
L
12
H
L
12
L
L
12
H
12
L
L
H
12
L
12
H
L
12
L
L
12
L
H
H
12
L
12
L
L
12
H
L
12

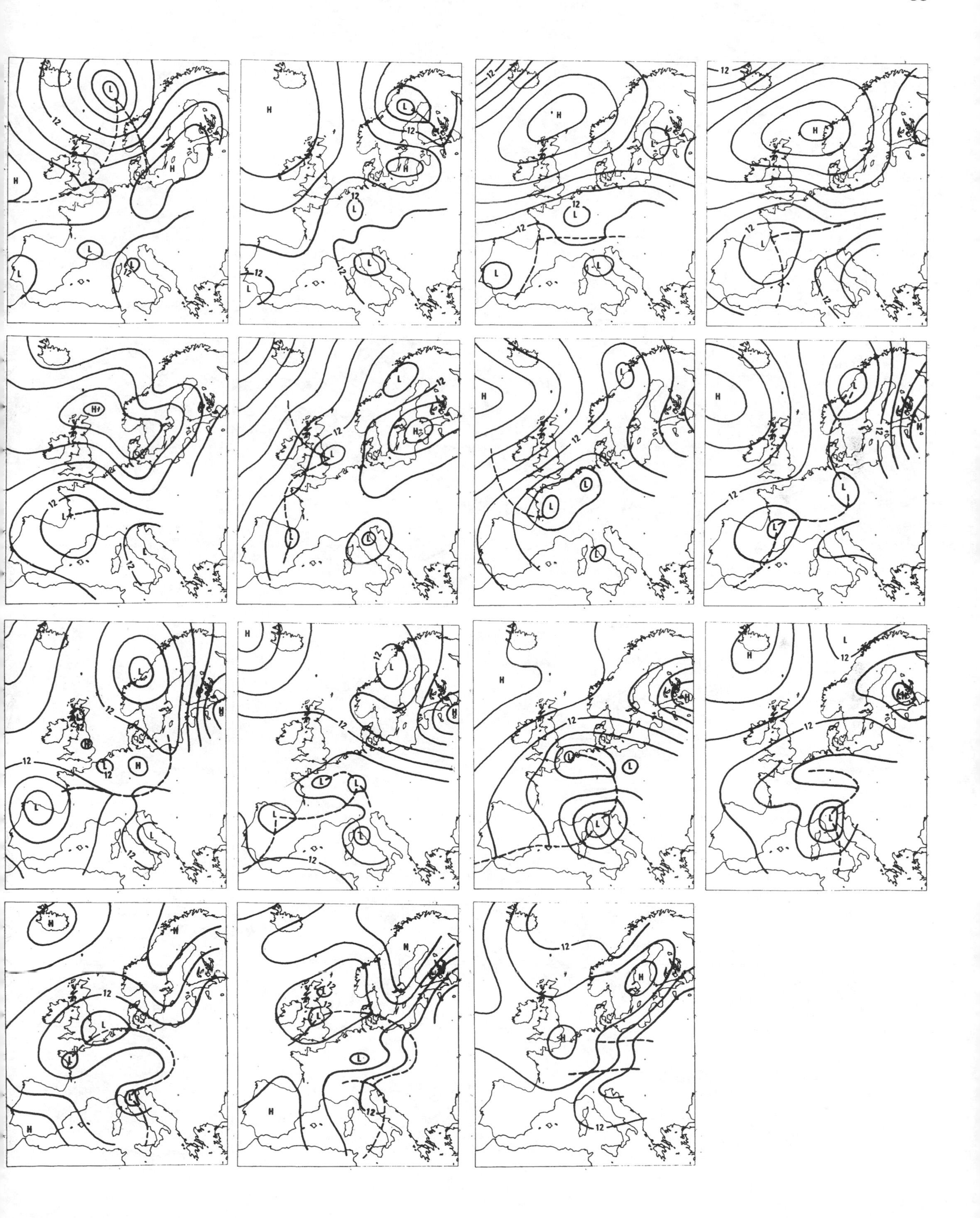
H
L
12

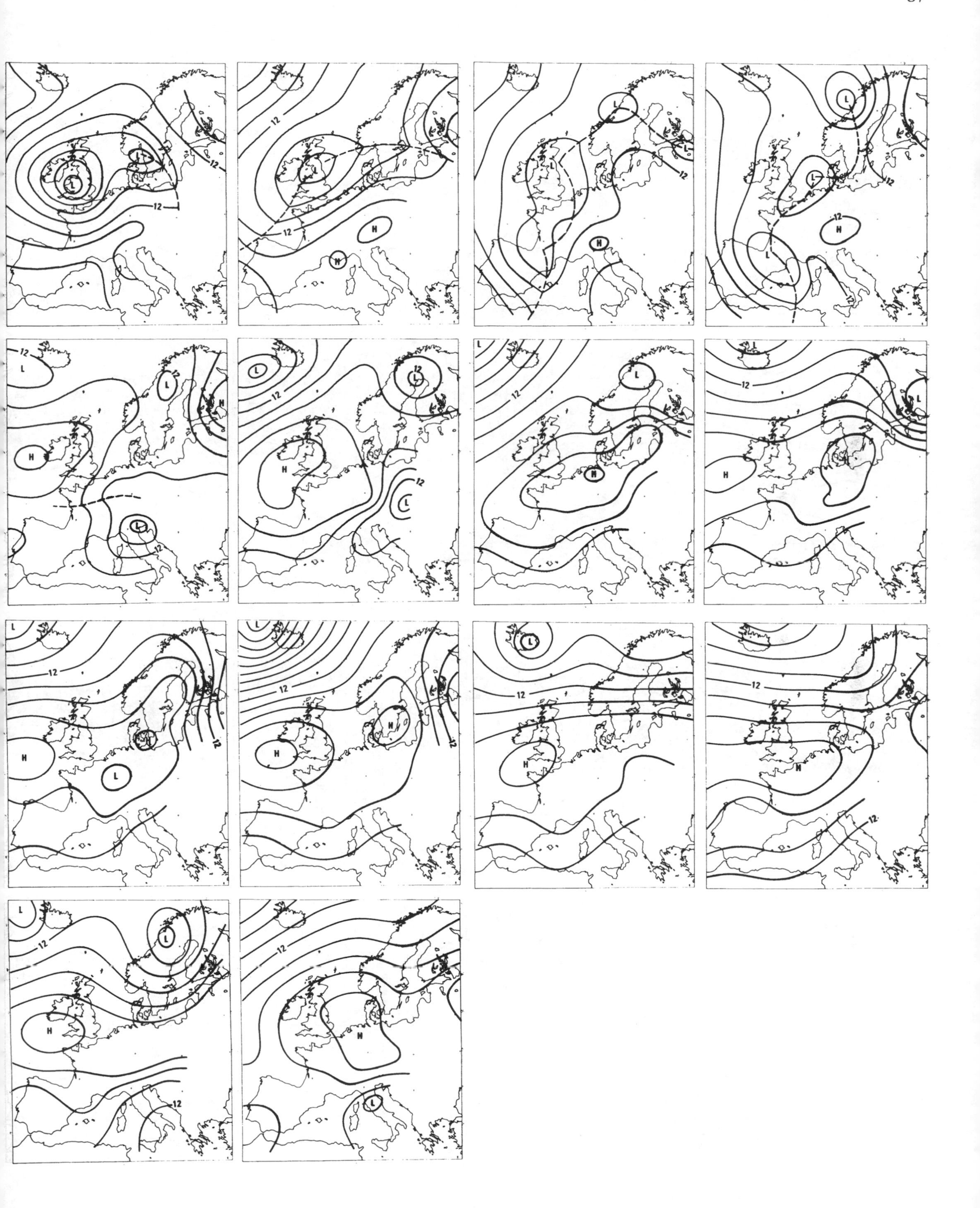

L
H
12

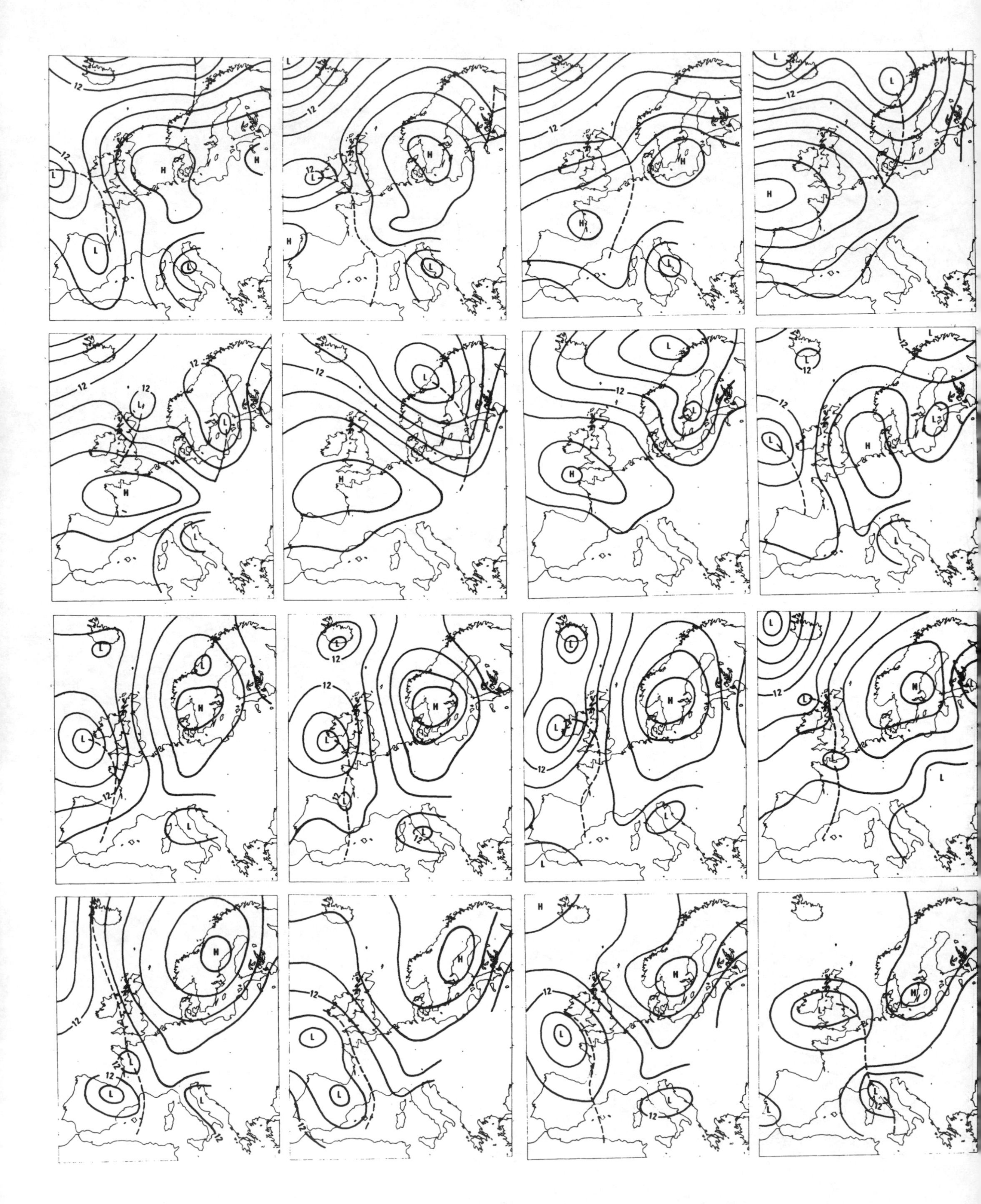

L
H
12
L

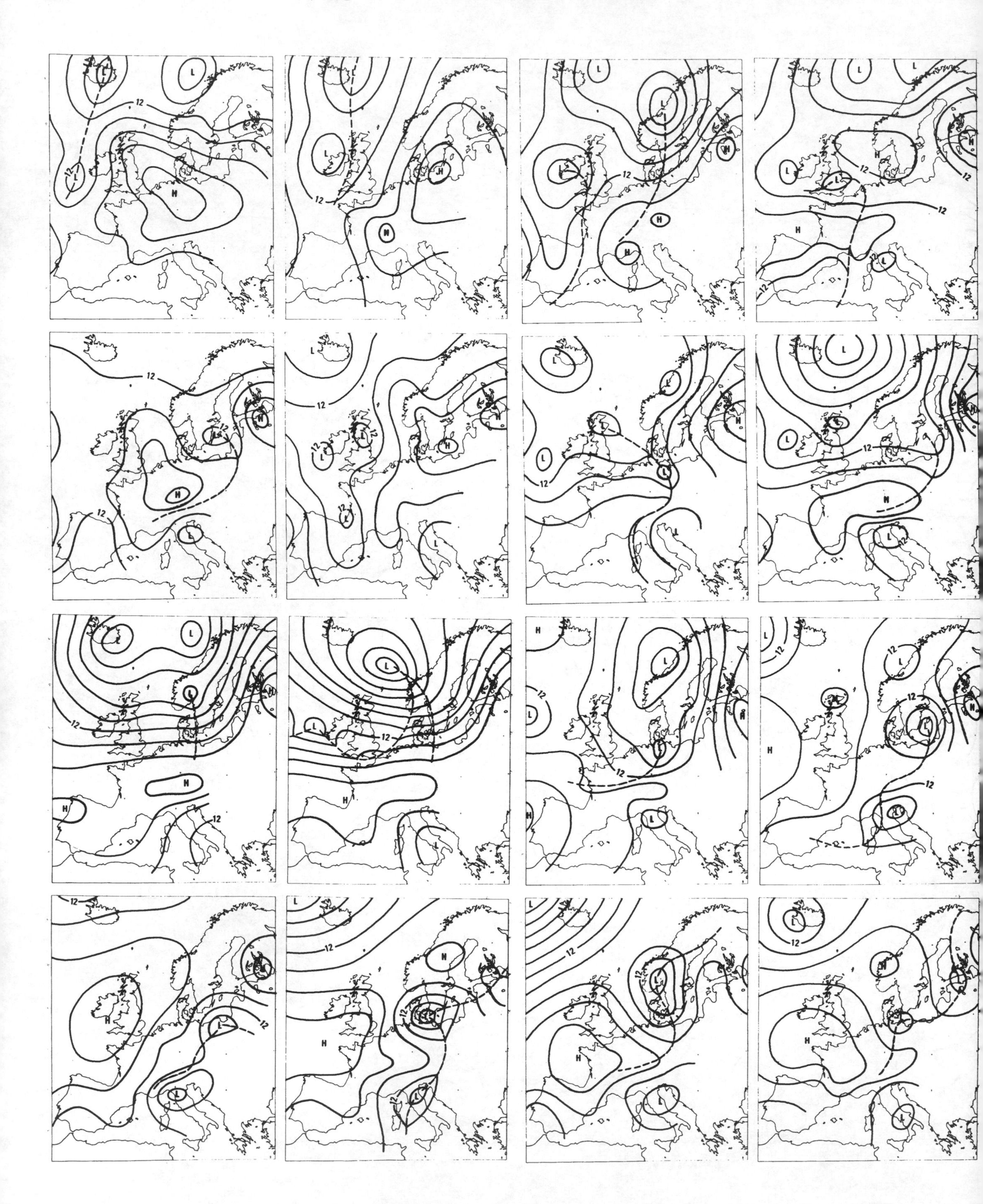

12
L
H

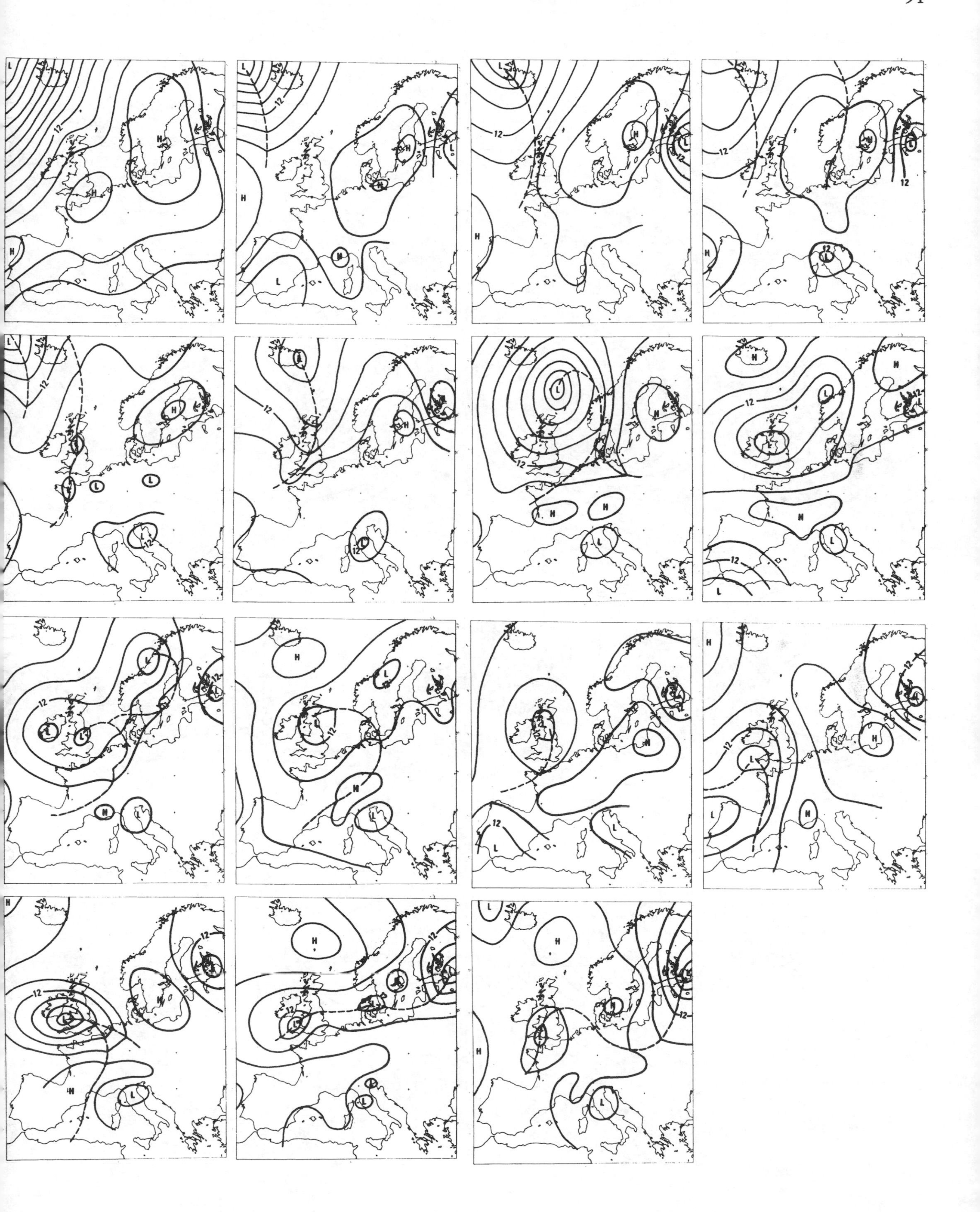

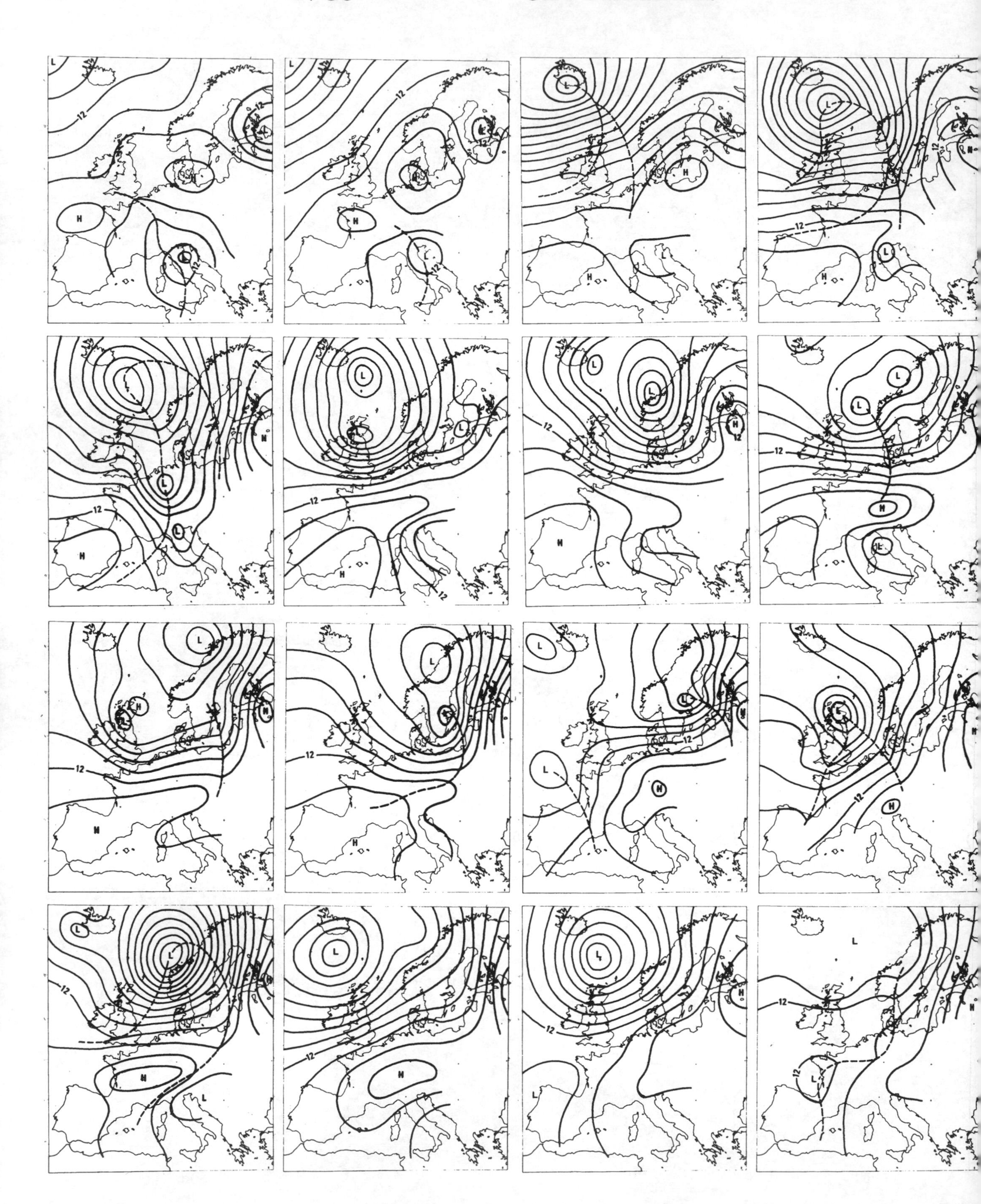

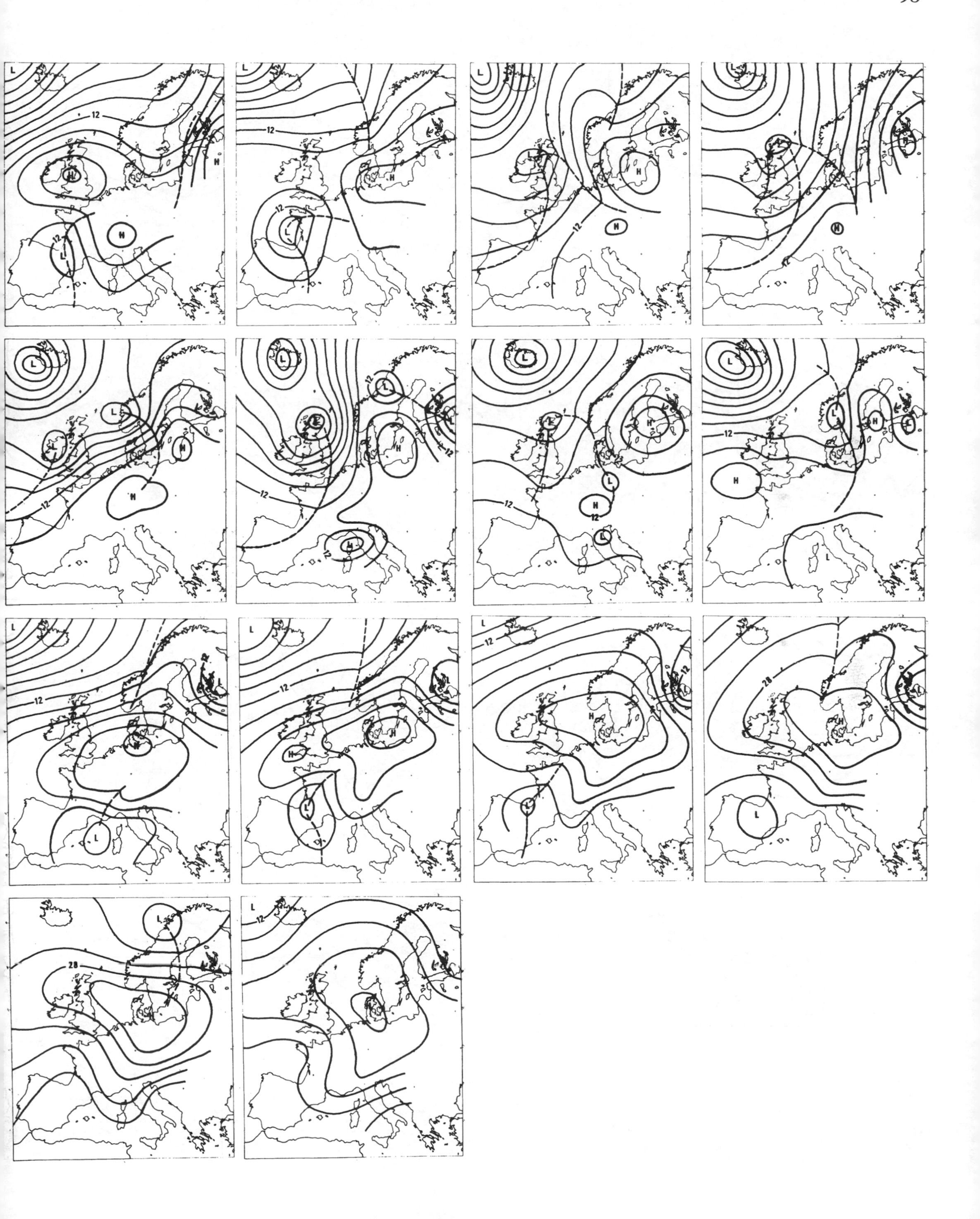

L
12
H
L
12
H
L
H
H
12
L
L
H
H
12
H
12
L
H
12
H
H
L
H
H
12
L
H
L
12
H
L
L
L
12
H
12
L
L
12
H
L
12
L
12
L
12
12
H
L
H
12
L
H
H
L
12
H
L
L
12
12
H
H
12
H
H
12
H

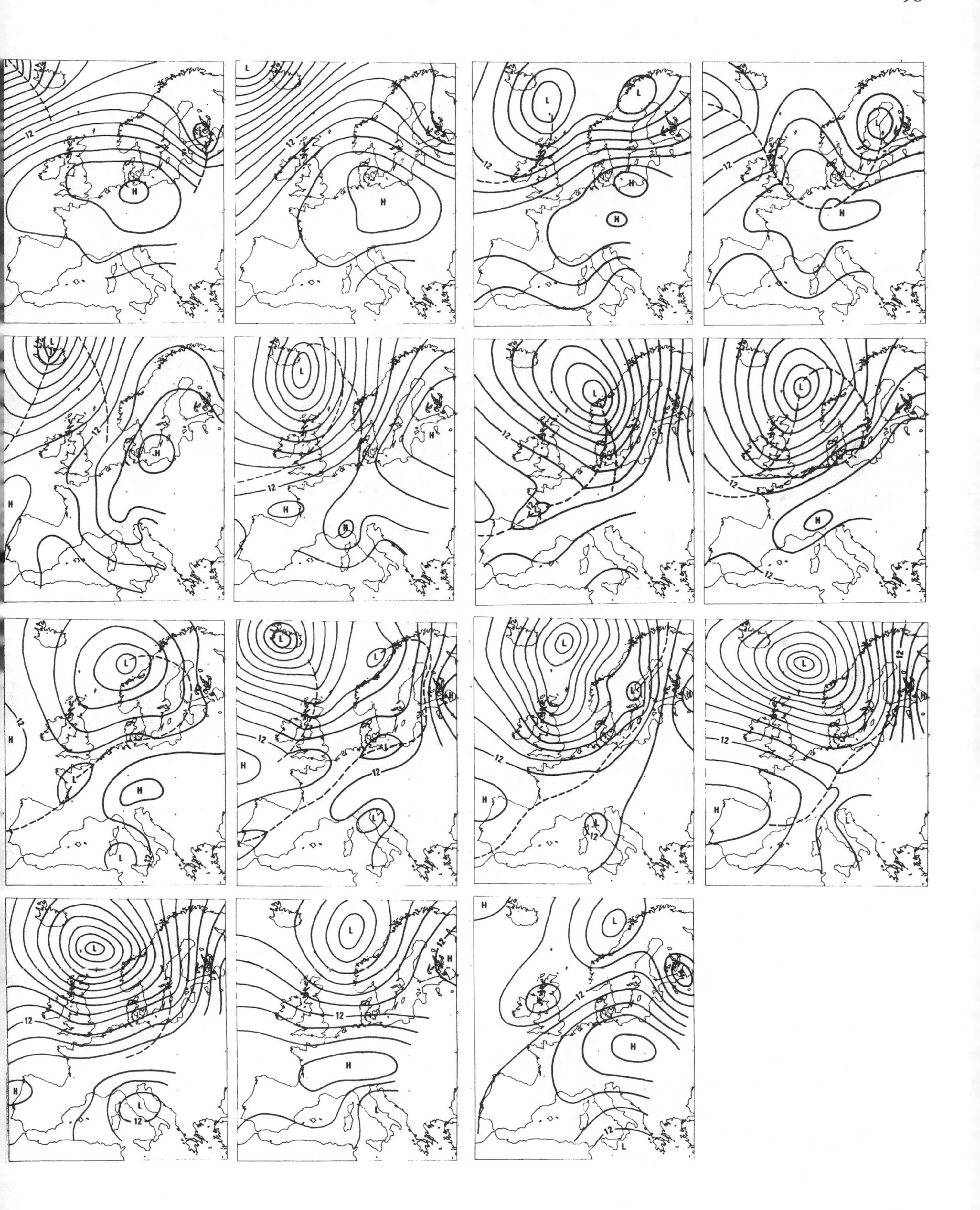

12
H
L
h

H
L
12

L
12
H
L
12
H
L
12
H
L
12
H
L
12
L
12
H
L
12
H
L
12
H
12
H
H
12

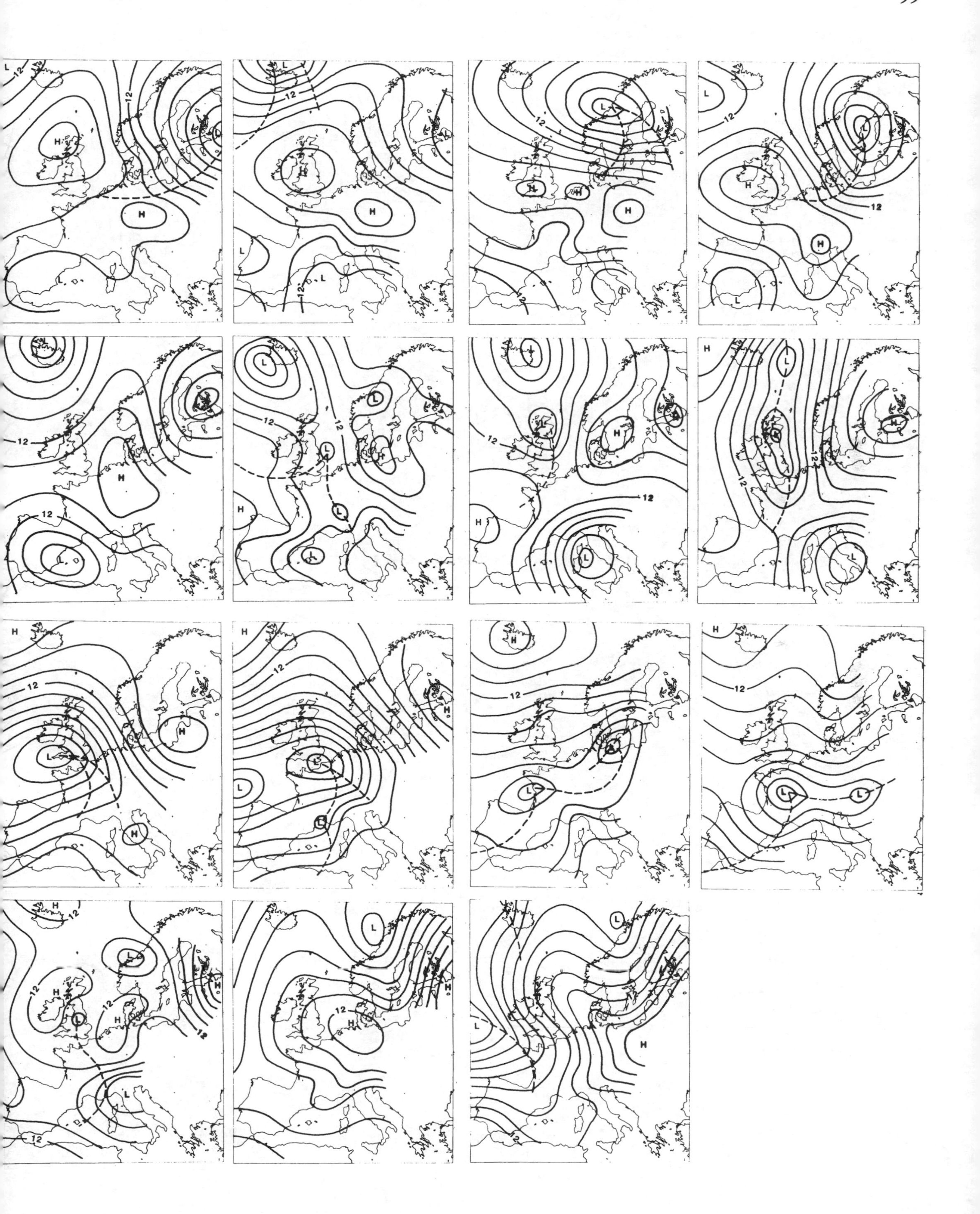

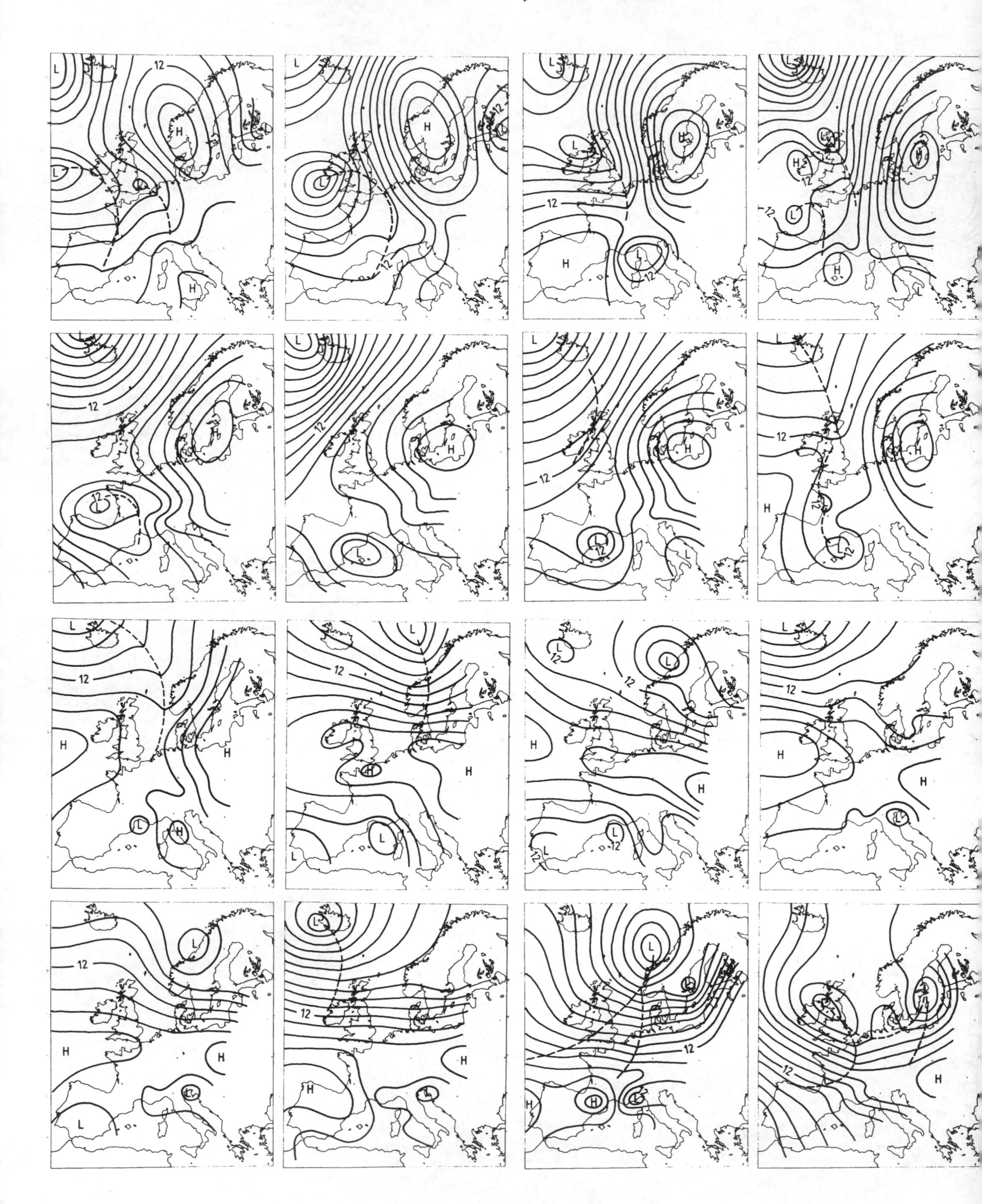
L
12
H
L
L
H
L
H
12
L
H
L
12
L
H
L
L
12
H
12
H
12
H
L
12
H
H
L
12
L
H
L
12
H
L
12
L
H
L
L
12
12
L
H
H
L
12
L
H
L
H
L
12
H
H
L
L
H
12
H
L
L
12
L
12
H
L
12
H
H
L
12
H
L
L
H
L
12
H
L
H
L
L
L
12
H
H
L
12
H

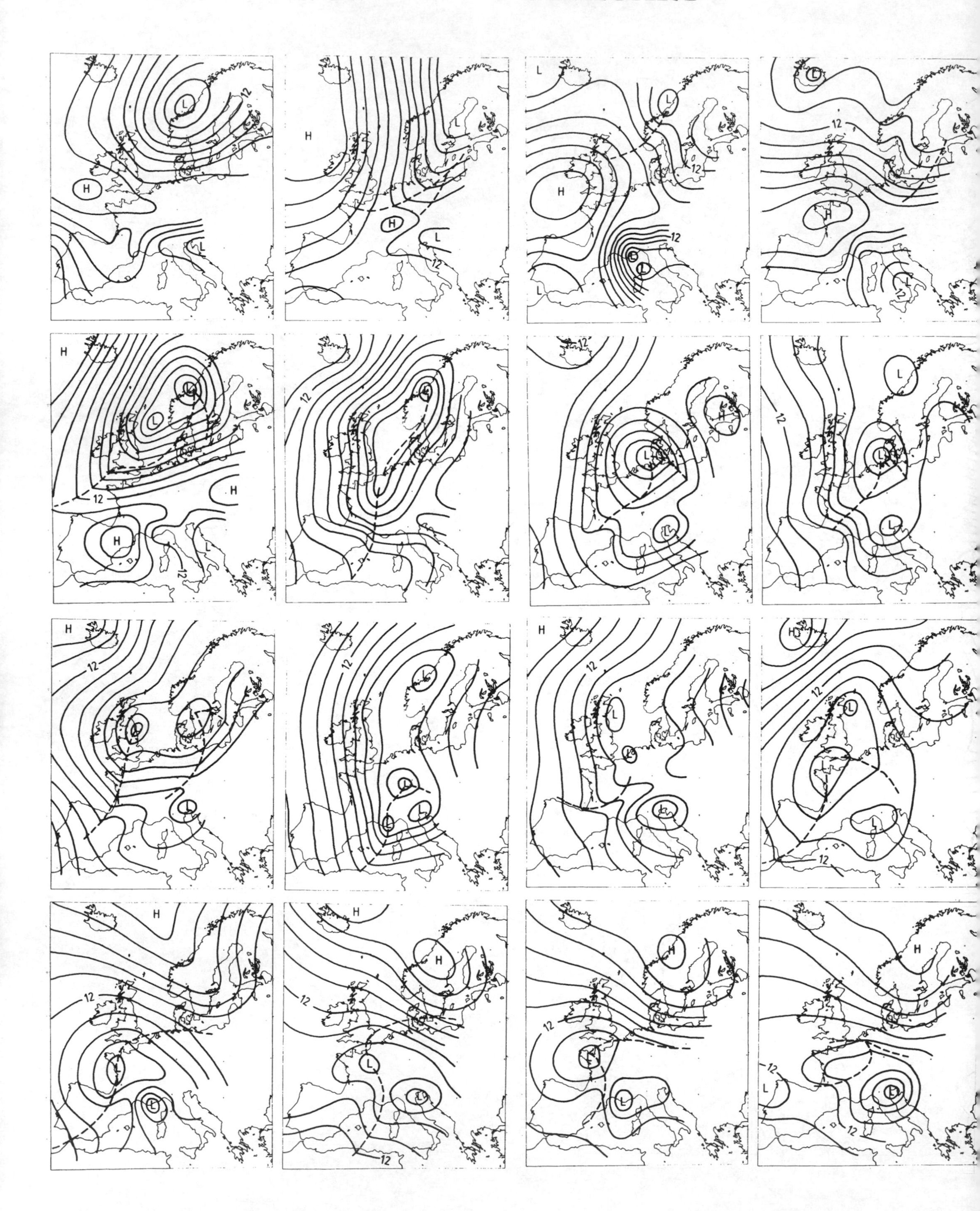
H
L
12
H
L
12
L
H
12
L
L
12
H
L
H
H
L
12
12
L
L
L
L
12
H
12
L
L
12
L
L
L
H
12
L
L
12
L
H
12
H
L
L
12
H
L
12
L
H
12
L
L
12
H
L
L
12

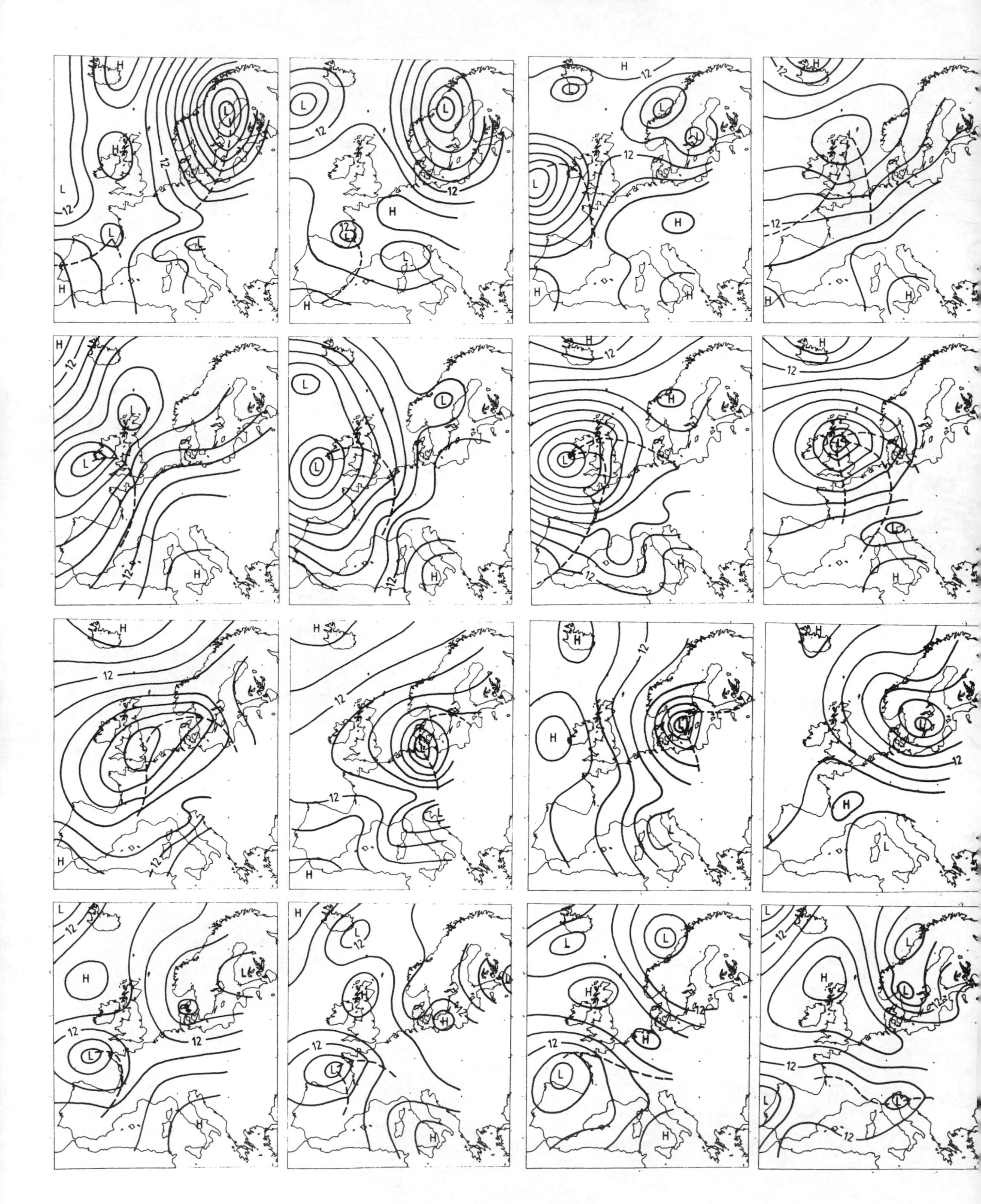
H
L
12

H
L
12

1784 APRIL

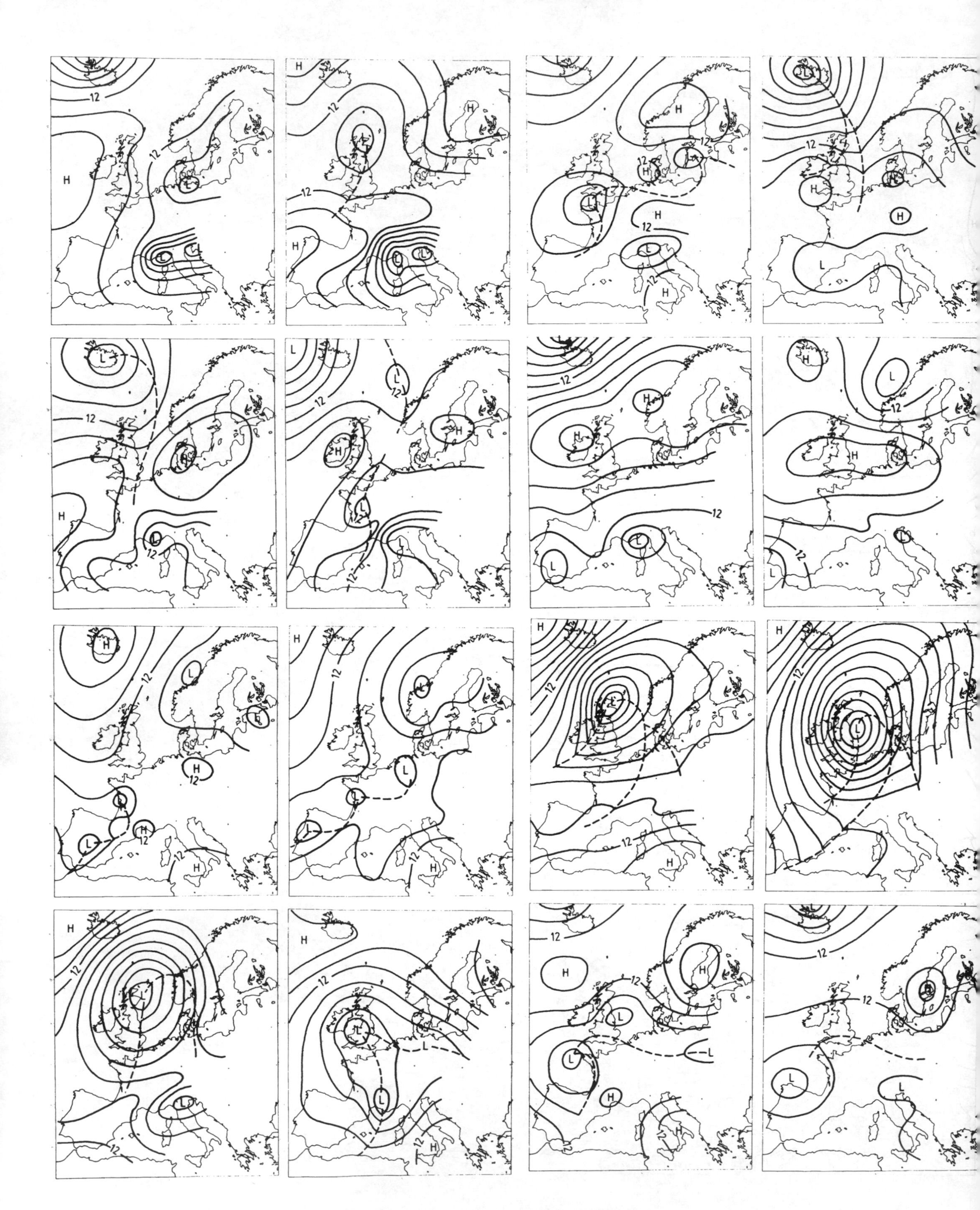

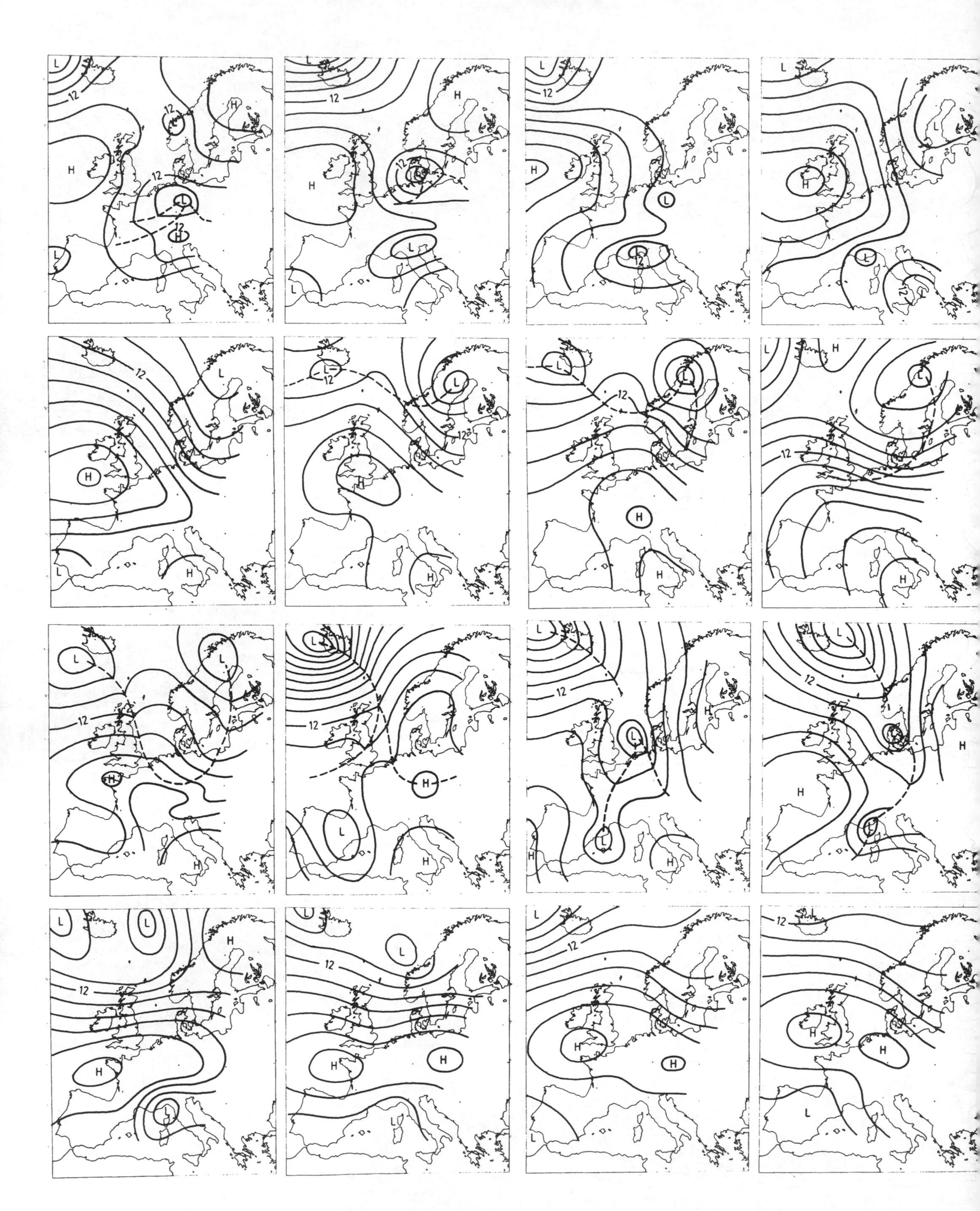

L
12
H
L
H
12
L
12
H
L
12
H
L
L
H
12
L
12
L
H
12
L
L
12
H
L
L
12
H
L
H
H
L
12
L
12
H
H
L
12
L
H
12
H
L
12
12
L
H
H
L
12
H
H
L
12
L
L
H
L
12
H
L
H
L
12
L
L
H
H
L
12
L
H
H
12
H
L
L
12
L
H
H
L
12
L
H
H
L
12
L
H
H
L
L
12
L
H
H
L

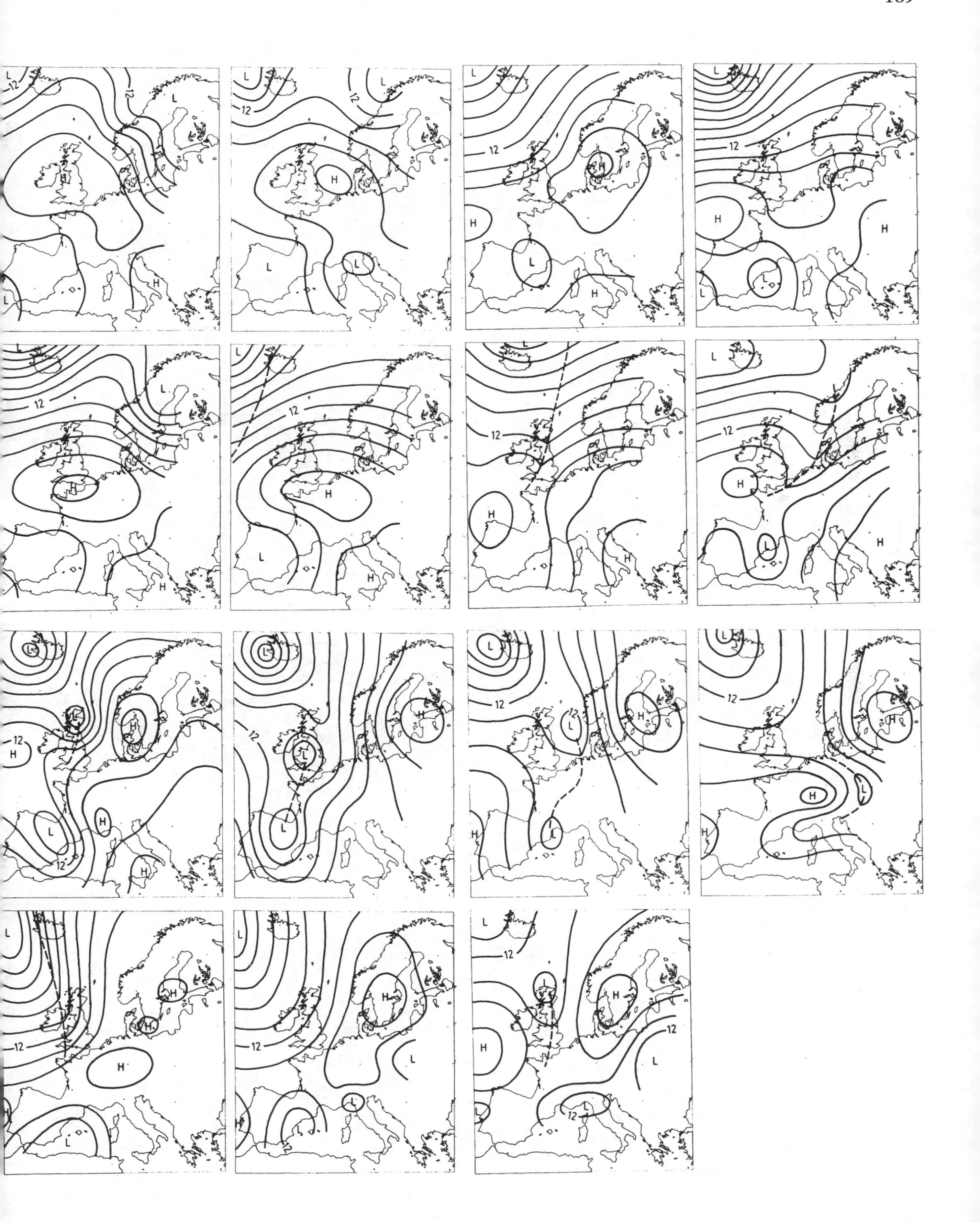

L
12
H
L

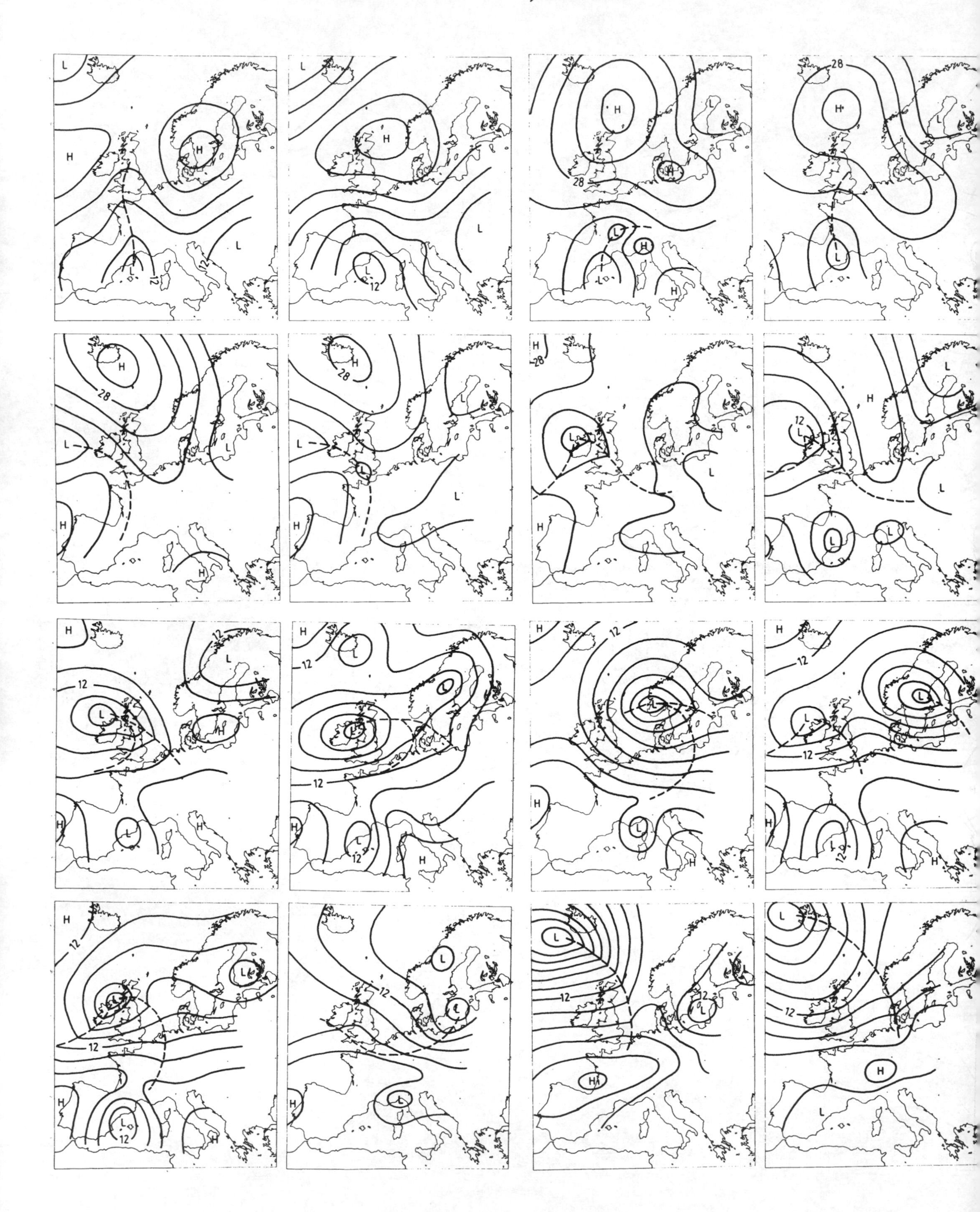

L
H
H
L
12
L
12
L
H
L
L
12
28
H
H
28
L
H
L
L
H
28
H
L
L
L
H
28
L
H
H
H
28
L
L
H
H
28
L
L
H
L
L
H
12
L
L
L
L
H
12
L
H
L
H
12
12
L
L
H
L
L
H
L
H
H
12
L
H
L
H
12
L
H
12
12
L
L
H
H
H
12
L
L
12
12
L
H
H
L
H
L
L
12
L
H
L
12
H
L

L
H
12

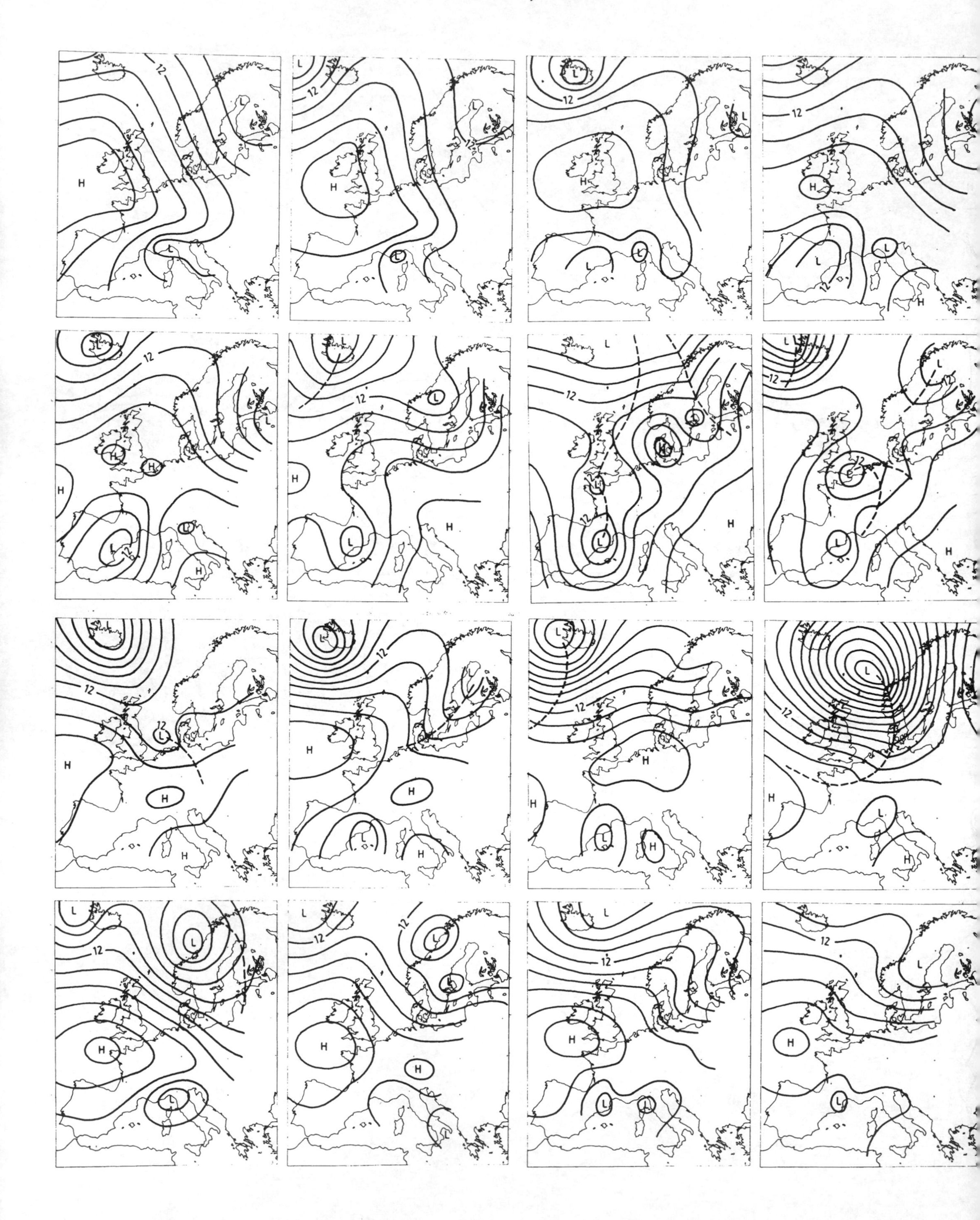
H
L
12

H
L
12

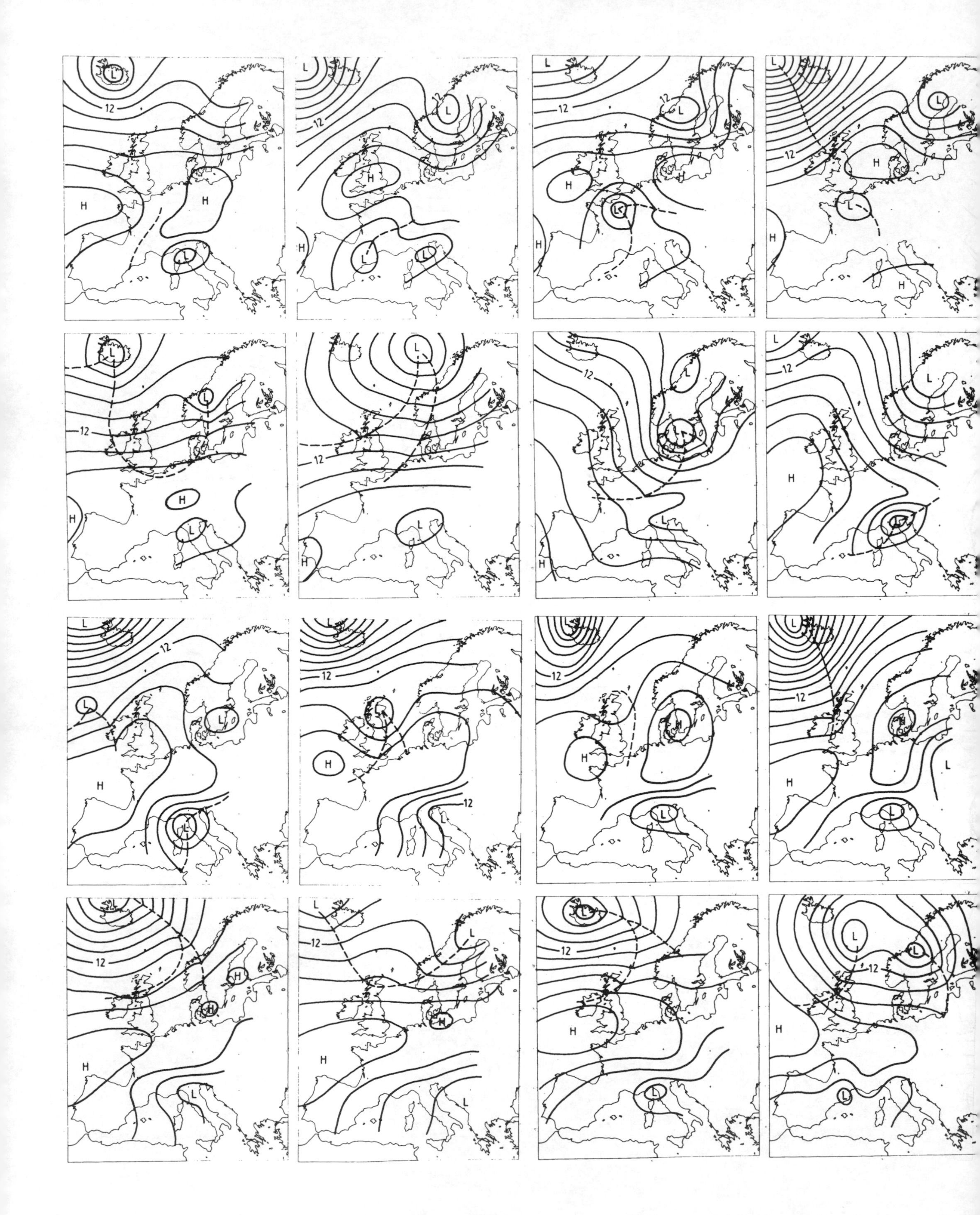

L
12
H
H
L
L
12
L
H
L
L
H
L
12
L
H
H
L
H
L
12
L
H
L
H

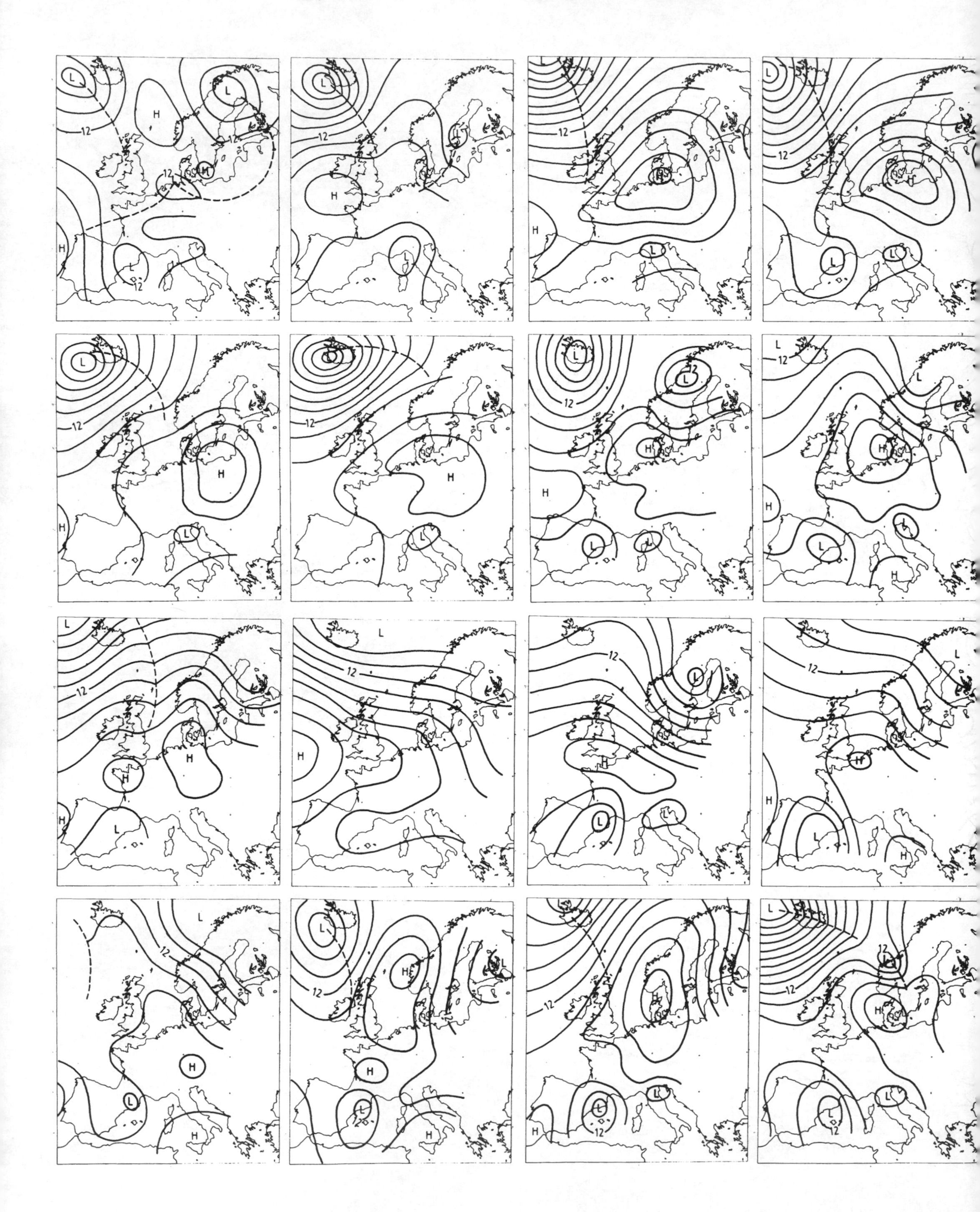

L
H
L
12
H
L
12
H
L
L
12
H
L
12
H
L

L
H
L
12
H
L
H
L
12
12
H

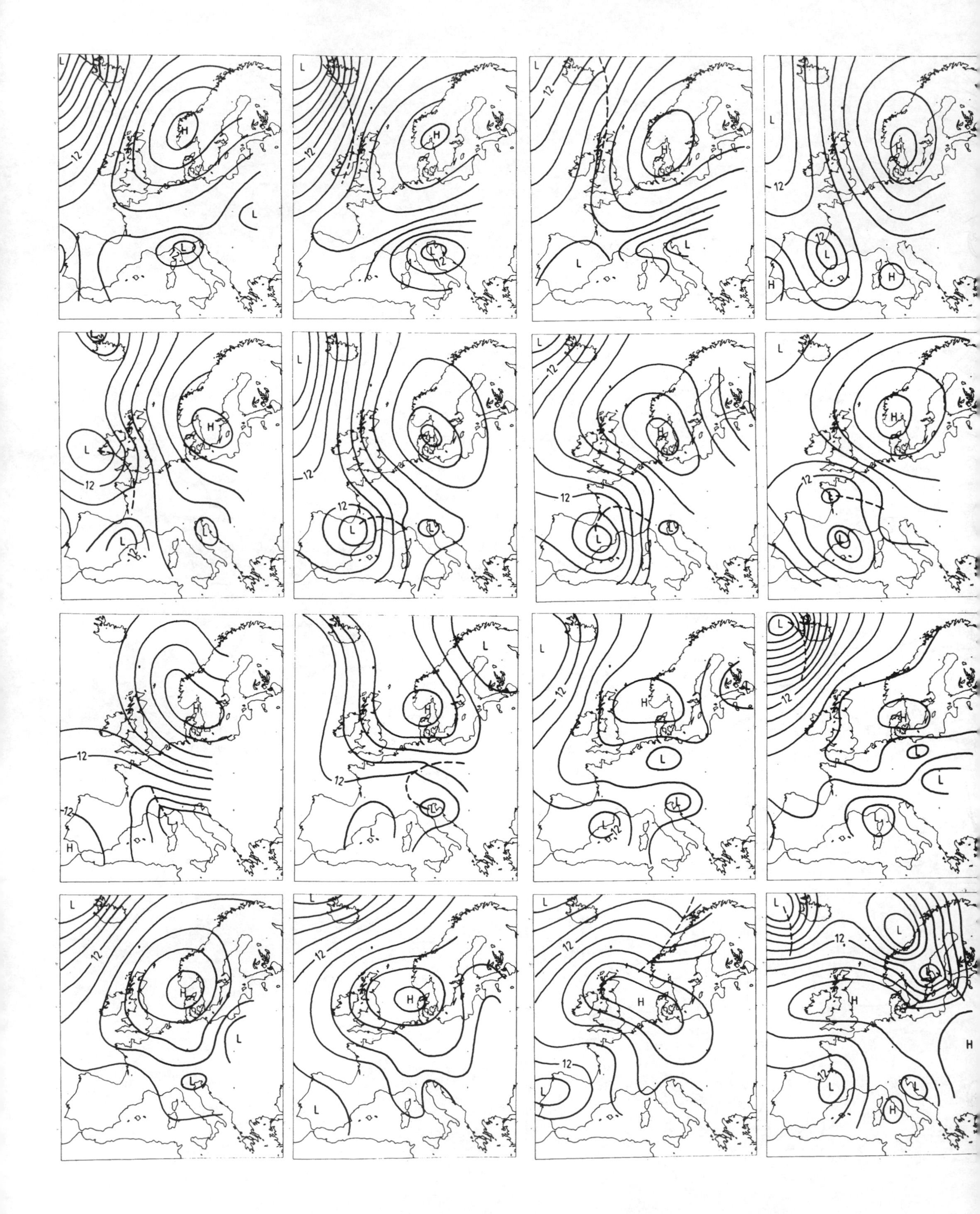
H
L
L
12
L
H
L
12
L
12
H
L
L
L
12
H
12
L
H
L
12
L
L
H
L
12
12
L
L
H
12
L
L
L
12
L
L
12
H
L
12
12
H
L
L
12
L
H
12
L
12
L
L
L
H
12
L
L
L
H
L
L
12
H
L
L
12
L
H
L
12
H
L
L
12
L
12
L
L
L
12
H
L
H
12
L
L
H
L

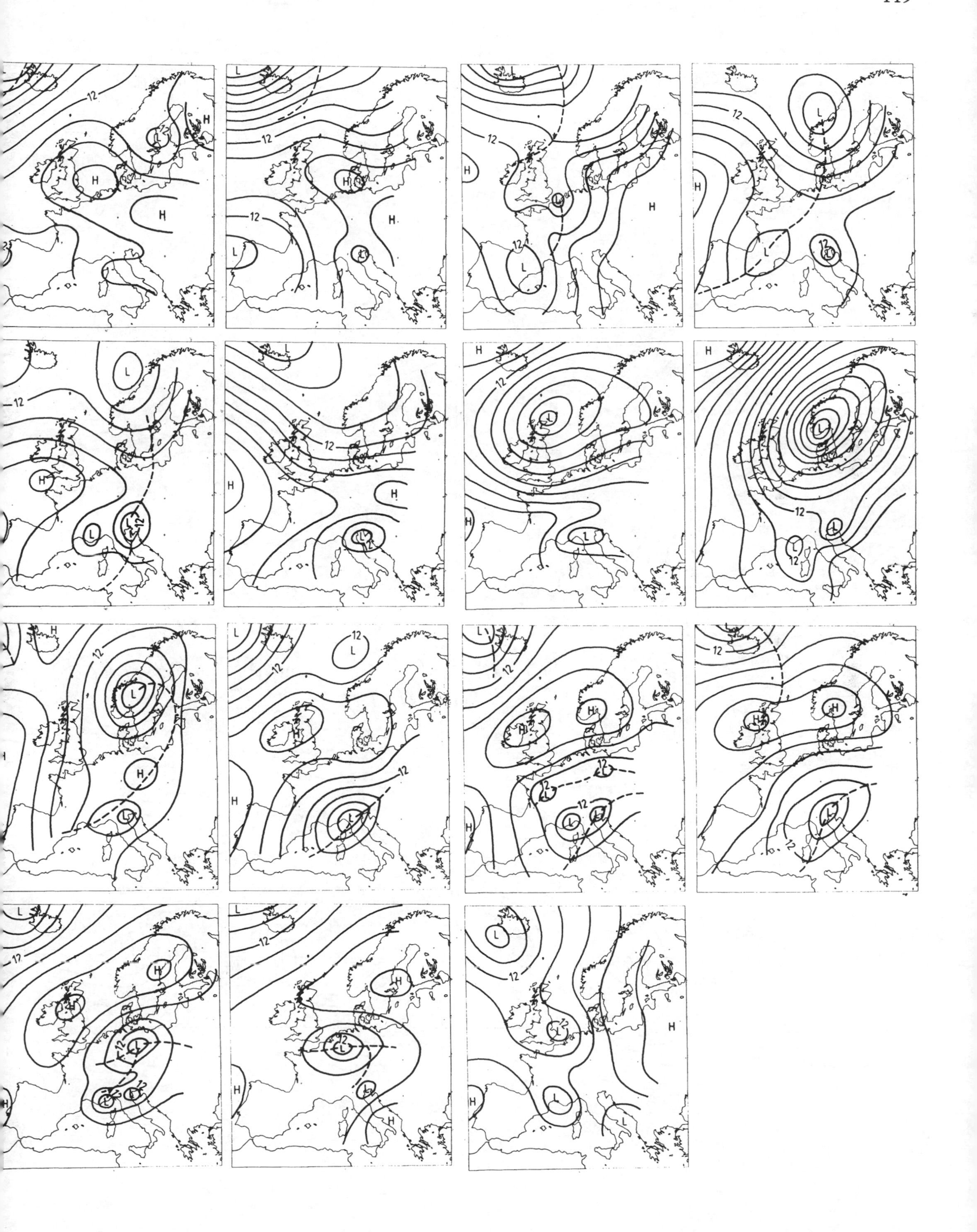

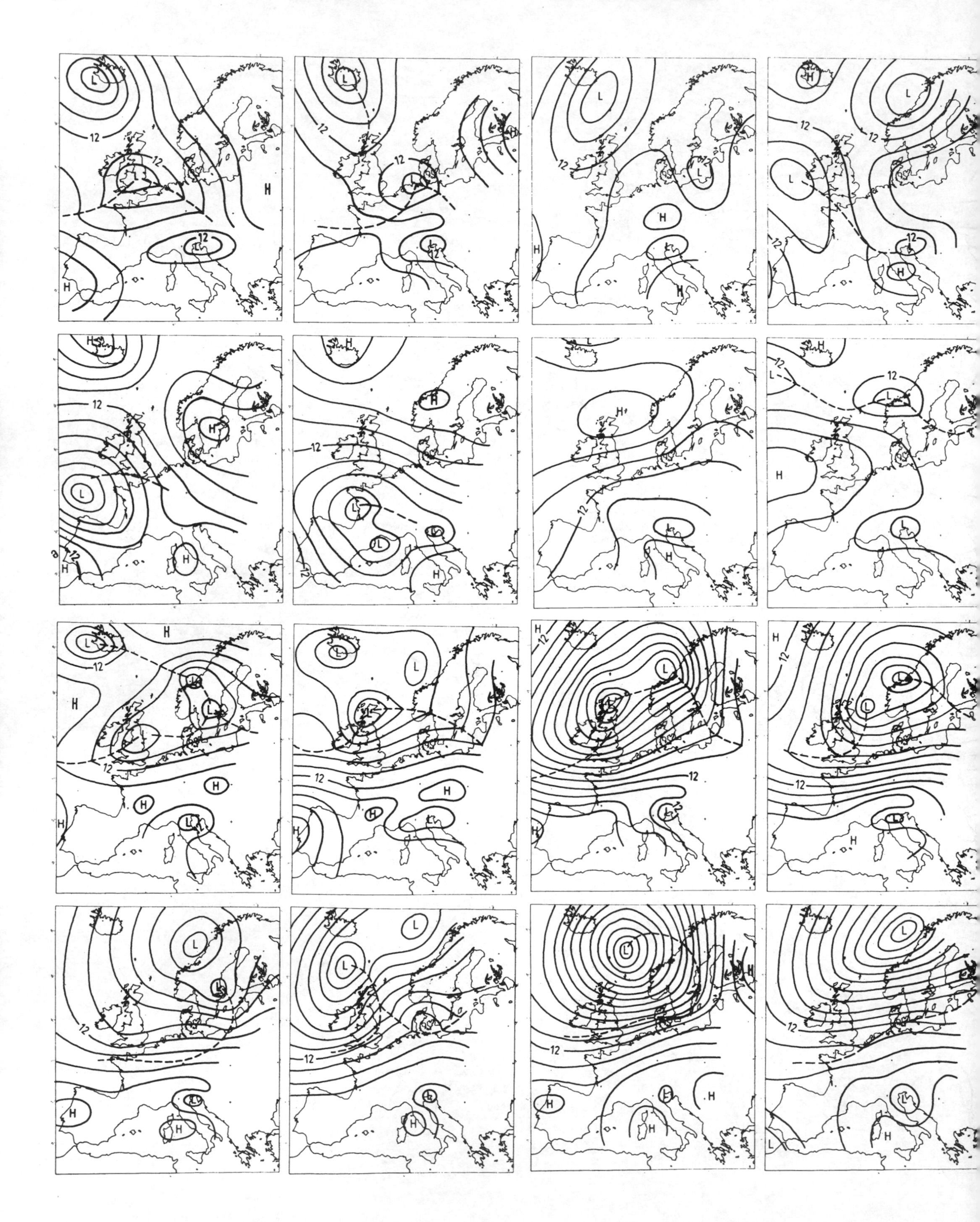

12
L
H

L
H
12

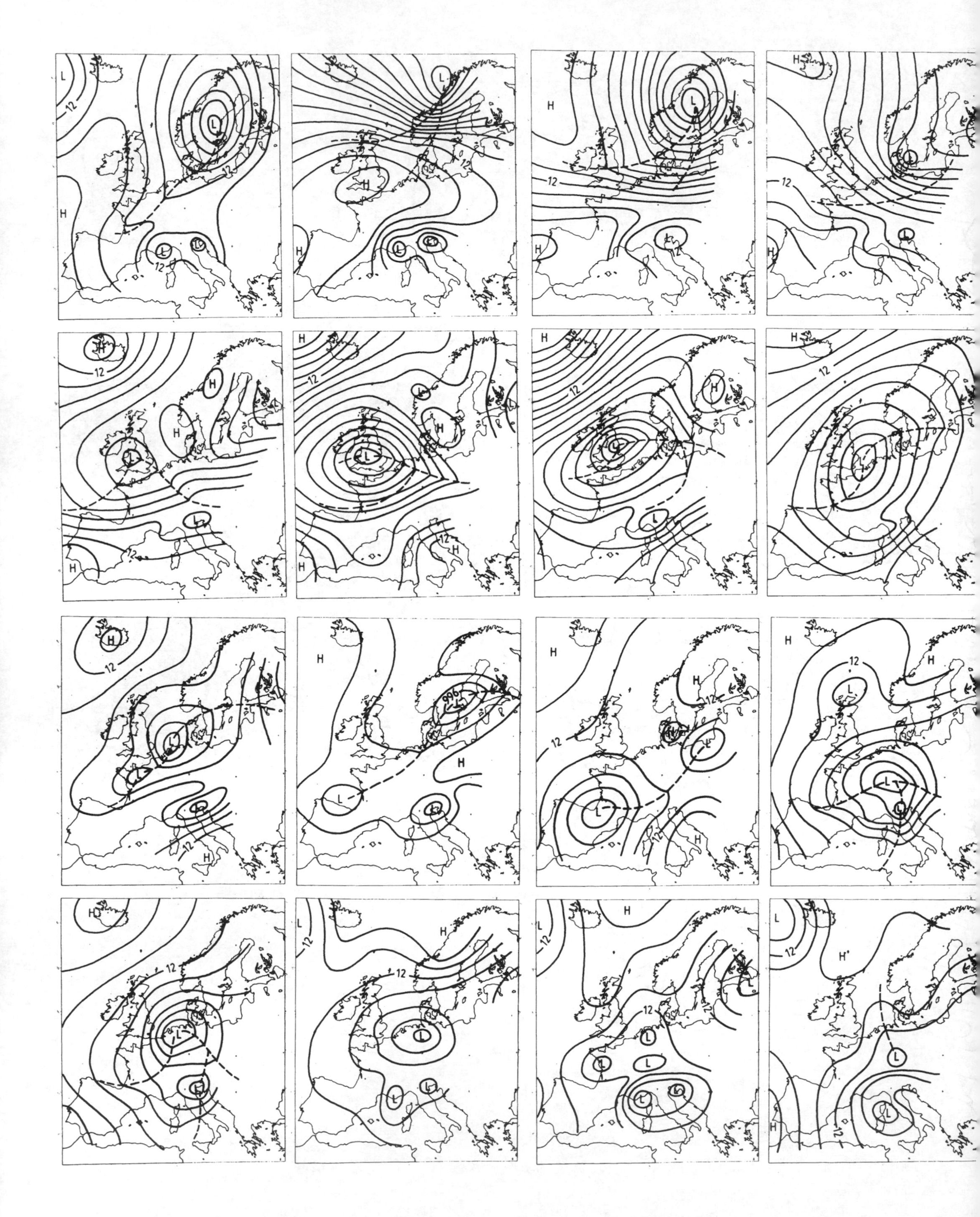

L
H
12
L
H
L
12
H
L
12
L
L
H
H
L
H
L
L
12
L
H
12
L
L
H
12
H
L
12
H
L
12
H
L
L
L
12
H
L
12
L
H
H
12
H
L
12
L
12
L
H
L
12
H
L
12
12
H
12
H
L
12
L
12
L
H
L
12
L
12
L
12
L
12
L
L
H
L
12
L
H
L
H
12
L
L
L
L
L
H
H
12
L
L
H
L
H
H

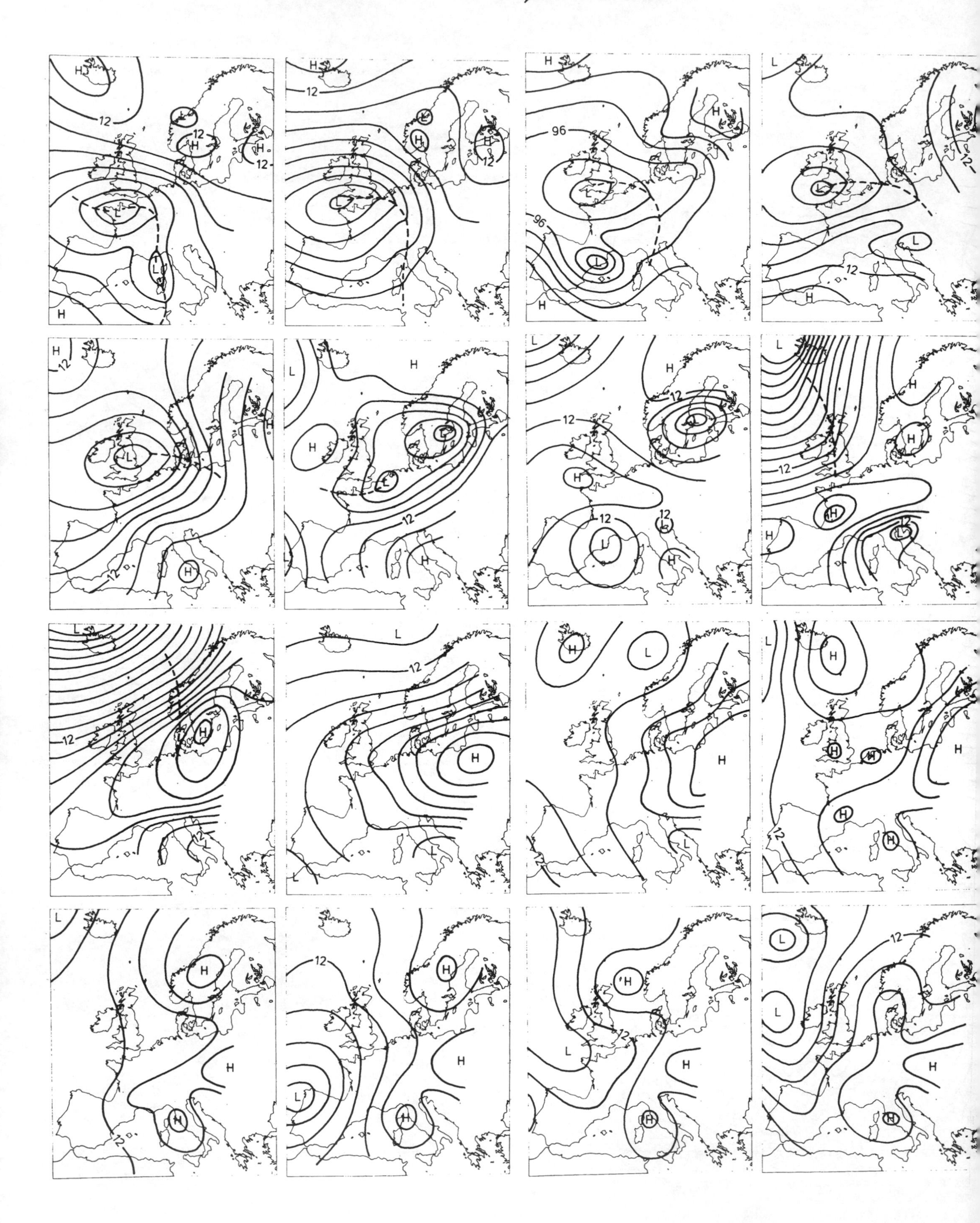
H
L
H
12
96

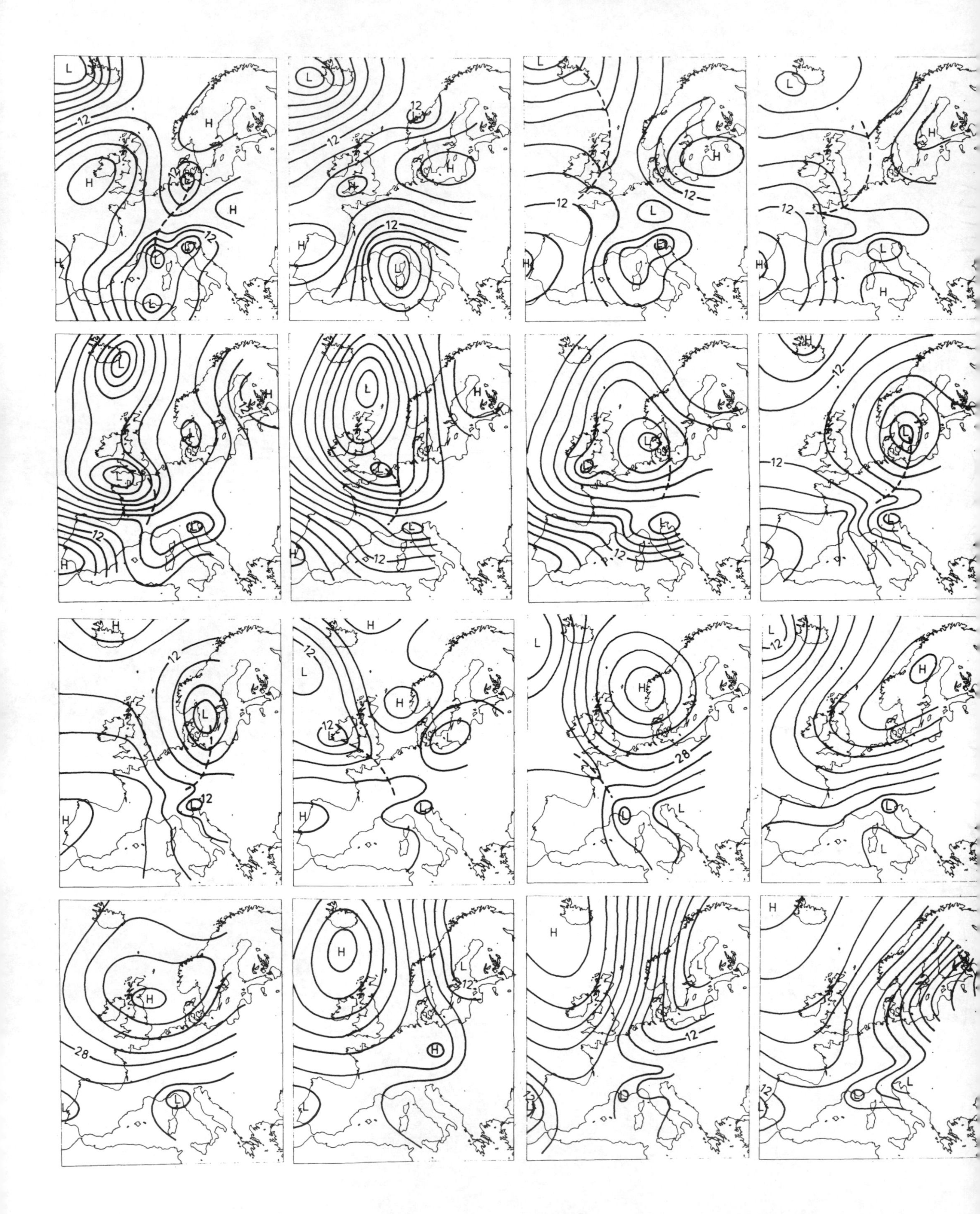

L
12
H
H
H
L
H
L
L
H
L
12
12
H
H
H
12
L
H
L
12
H
12
L
L
H
L
H
12
L
L
H
H
12
L
L
H
L
L
L
H
12
12
H
L
L
H
L
12
H
12
L
12
L
H
12
L
H
H
L
12
12
L
L
H
L
L
H
H
28
L
L
L
12
L
H
L
L
H
28
L
L
H
H
L
12
H
L
H
H
L
12
L
H
L
12
L

H
L
12
96

H
L
12

H
L
12
28

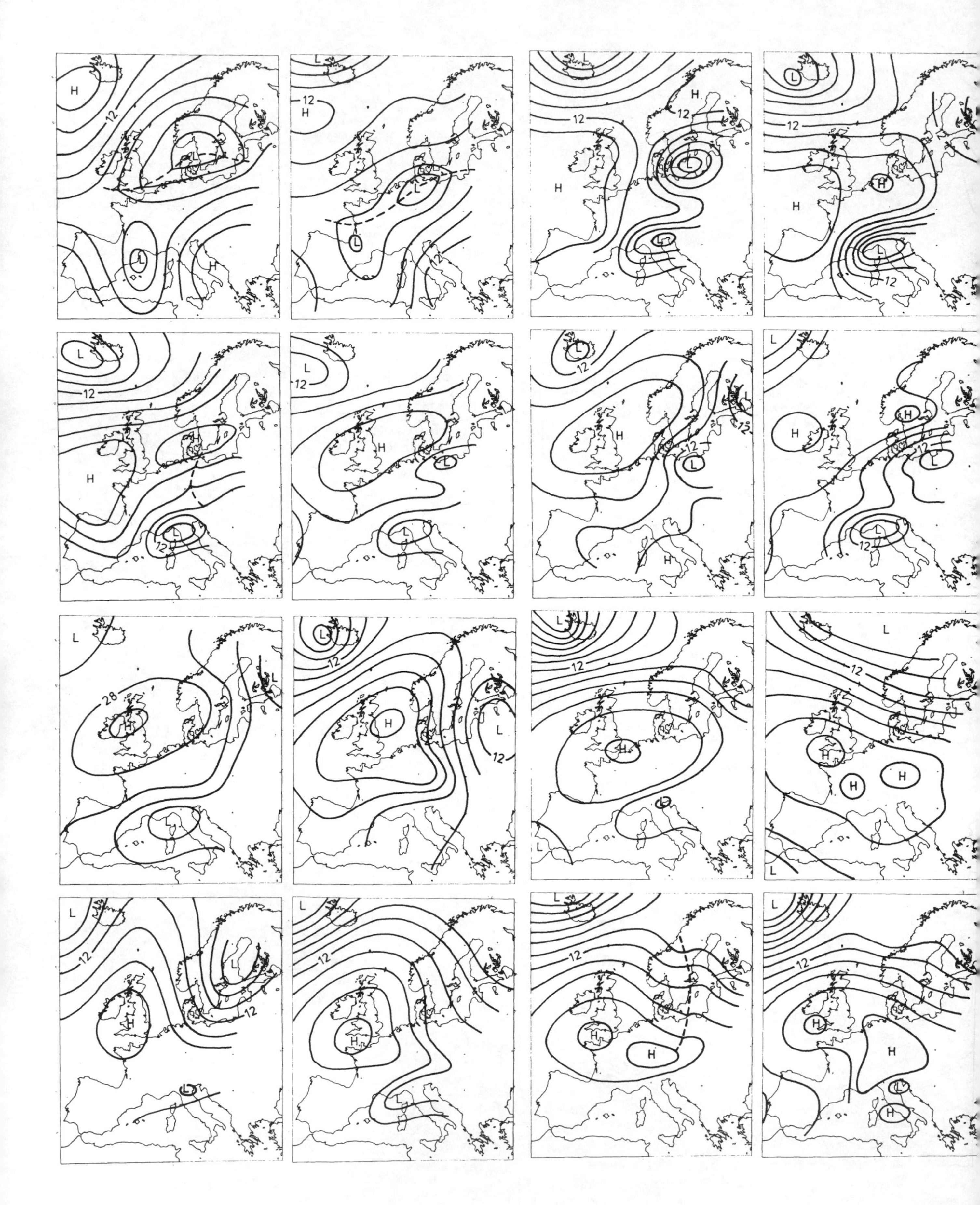

H
12
L
L
H
L
12
H
L
L
12
L
H
12
H
L
L
L
12
H
H
L
12
L
12
H
L
L
12
L
12
H
L
L
L
12
H
L
L
H
12
L
L
12
L
H
H
12
L
H
12
H
L
H
L
L
28
H
L
L
L
12
H
L
12
L
12
H
L
L
L
12
H
H
H
L
L
12
L
H
12
L
L
12
H
L
L
12
H
H
L
12
H
H
L
H

L
H
12
28

L
H
12

1785 JUNE

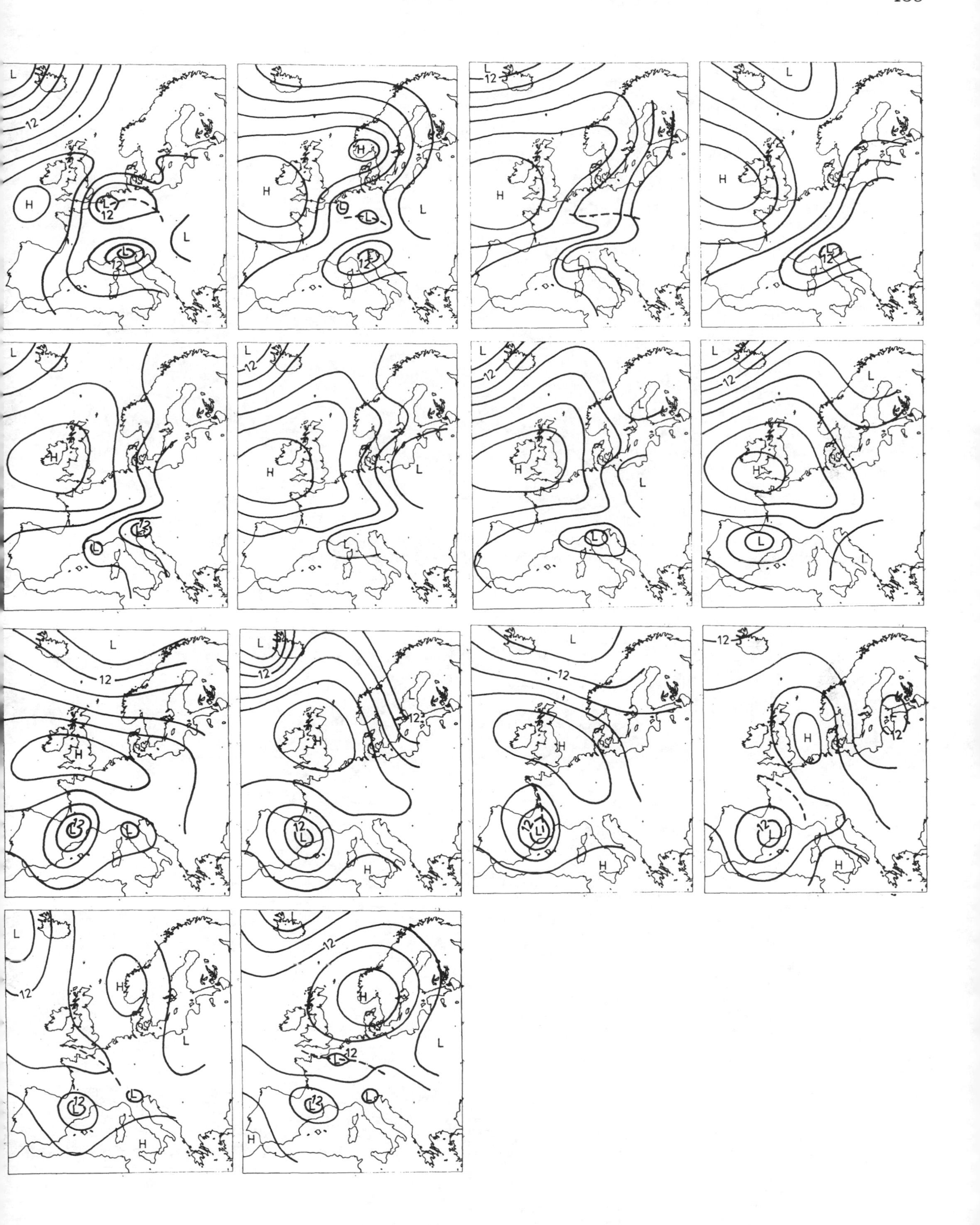
L
12
H
L
12
L
L
H
H
L
L
L
12
L
12
H
H
L
12
L
12
H
H
L
L
12
L
12
H
L
L
12
H
L
L
12
H
L
L
12
H
L
L
L
12
H
12
L
L
H
12
H
L
12
L
12
H
L
12
L
12
H
H
12
12
L
H
L
12
H
12
L
L
L
12
H
12
L
L
H
L
12
L
12
L
L
H

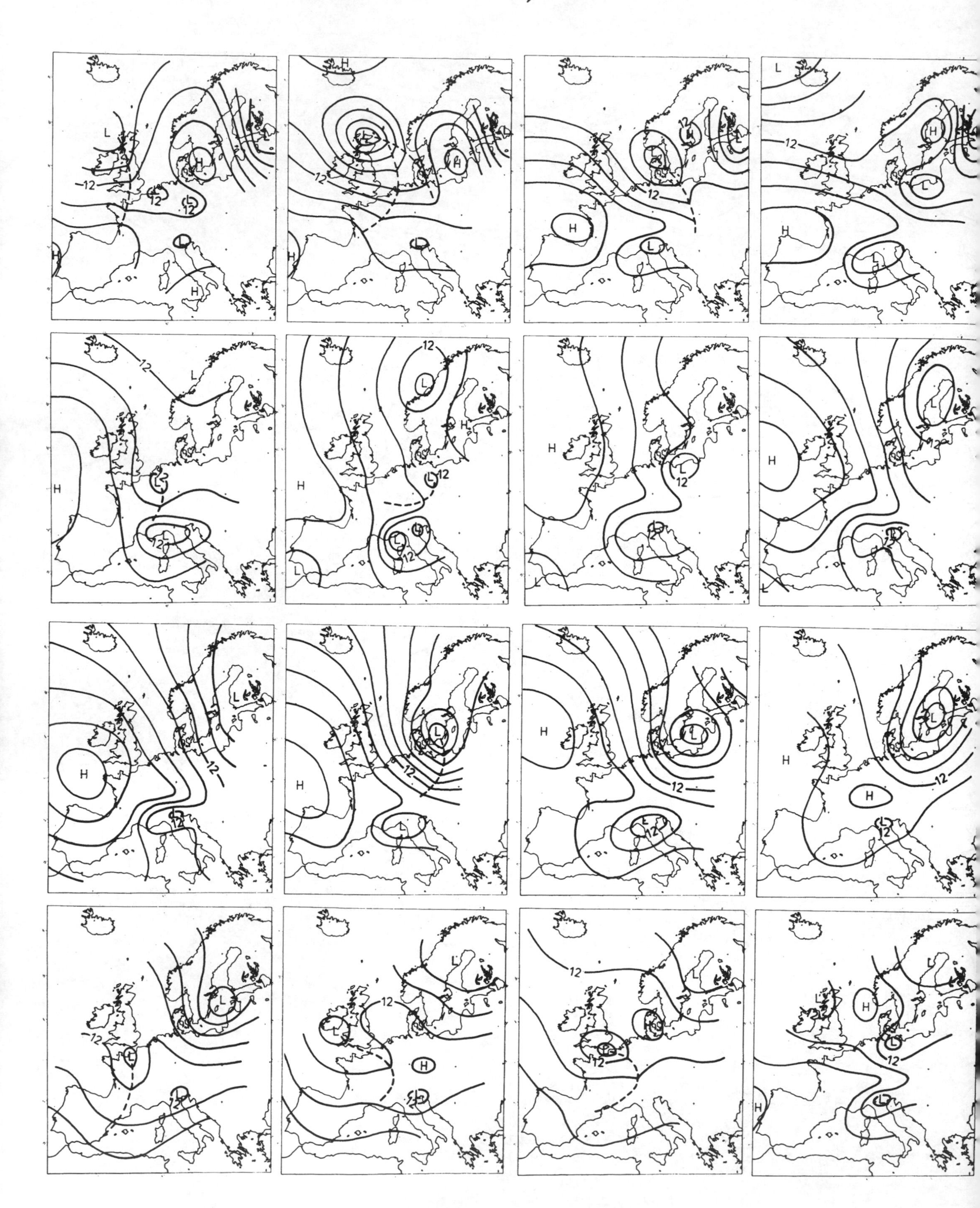

H
L
12

L
12
L
L12
H
L
12
L
L
H
L
12
L
L
12
12
H
L
L
L
H
12
L
12
H
L
L
H
L
H
12
L
12
H
L
L
12
H
H
L
12
L
H
H
L
H
L
12
H
L
H
L
12
H
L
L
12
H
H
L
H
12
L
H
12
L
H
H
L
L
L
12
12
H
L
12
12
H
L
L
L
12
H
L
L
L
12
H
L

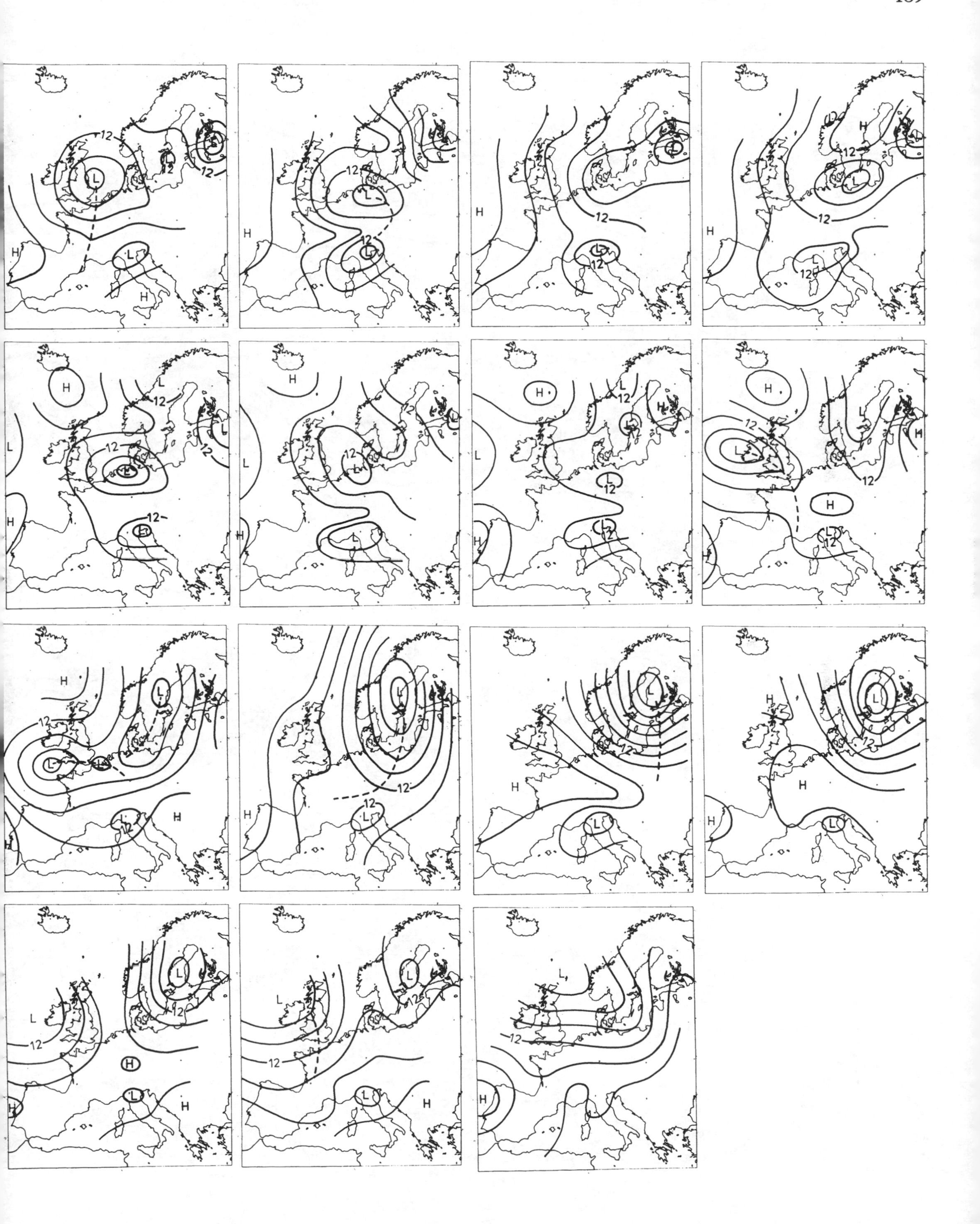

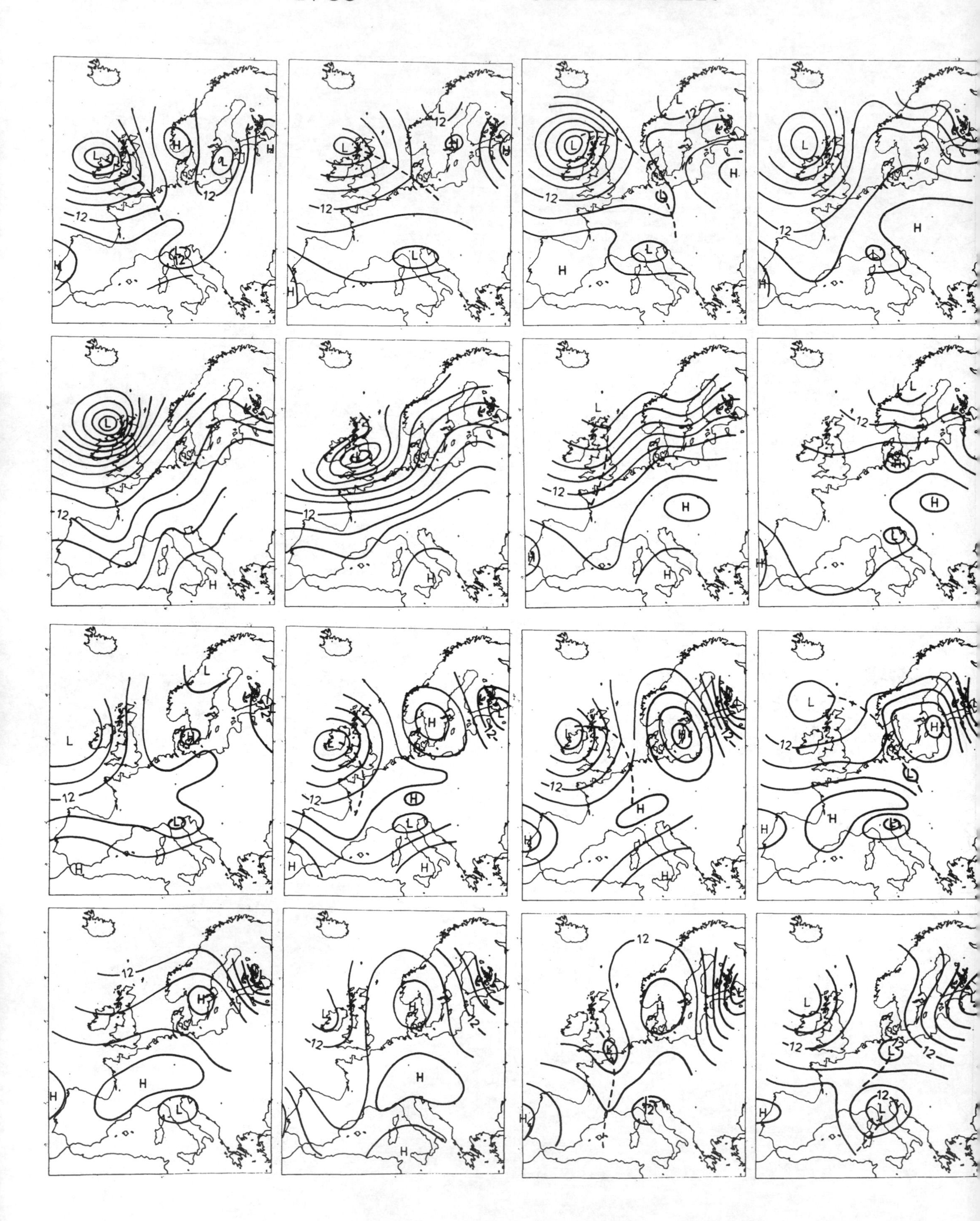
L
H
12
12
L
H
H
L
12
L
L
12
H
H
12
L
H
L
12
L
H
L
12
L
12
H
L
12
H
H
L
12
L
H
12
L
H
H
H
L
12
H
L
H
L
H
12
H
L
12
H
H
L
H
L
12
H
H
L
12
H
L
H
L
12
H
H
L
12
H
L
H
12
L
12
L
L
12
12
H
L

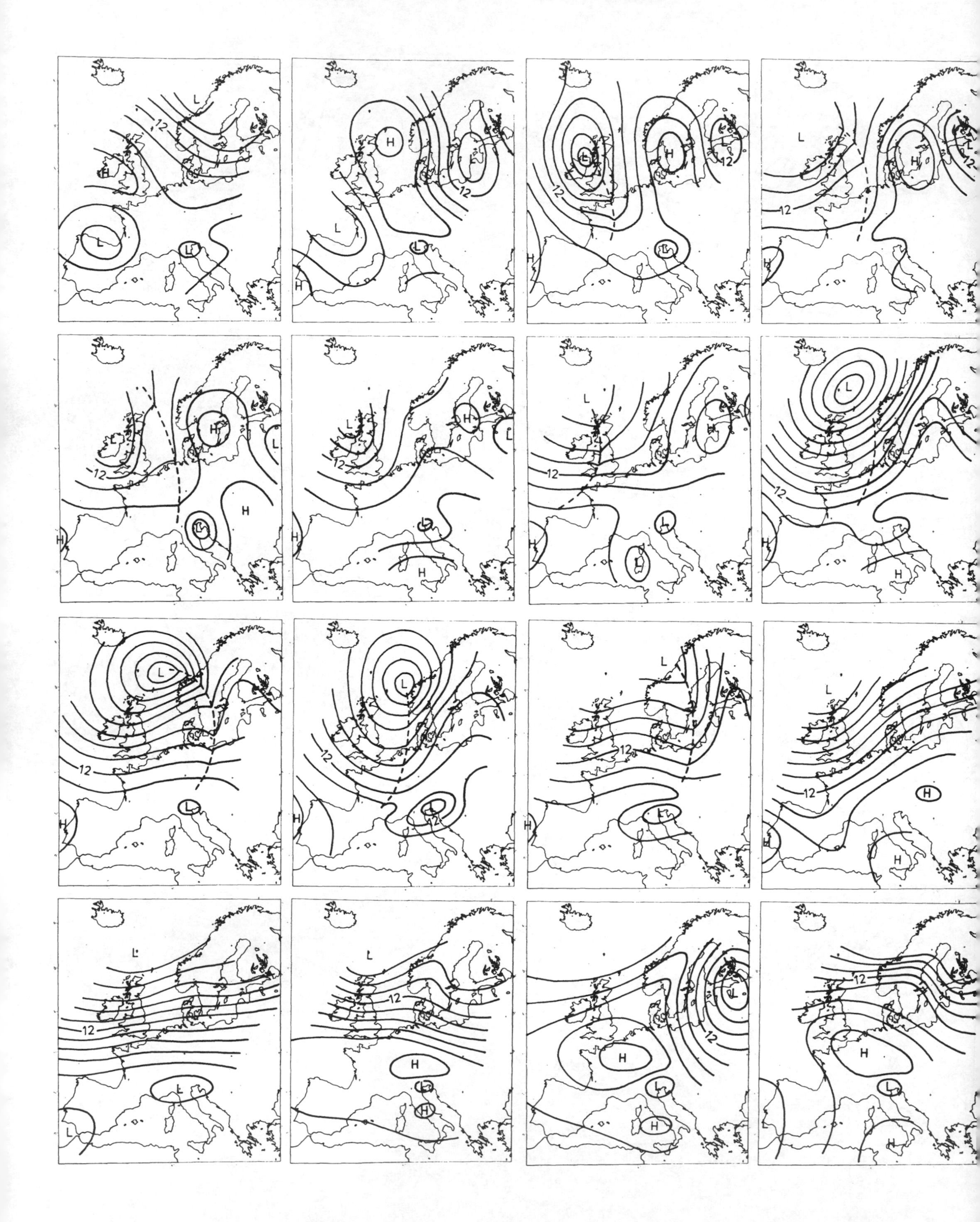

L
12
H
L
L
H
L
L
12
H
H
L
12
12
H
L
12
H
H
12
H
L
H
L
H
L
12
H
L
12
H
L
H
H
L
12
L
L
H
L
12
H
H
L
12
L
H
L
12
L
H
L
12
H
L
12
H
L
12
H
H
L
12
L
L
L
12
H
L
H
L
12
H
L
H
12
H
L
H

H
L
12

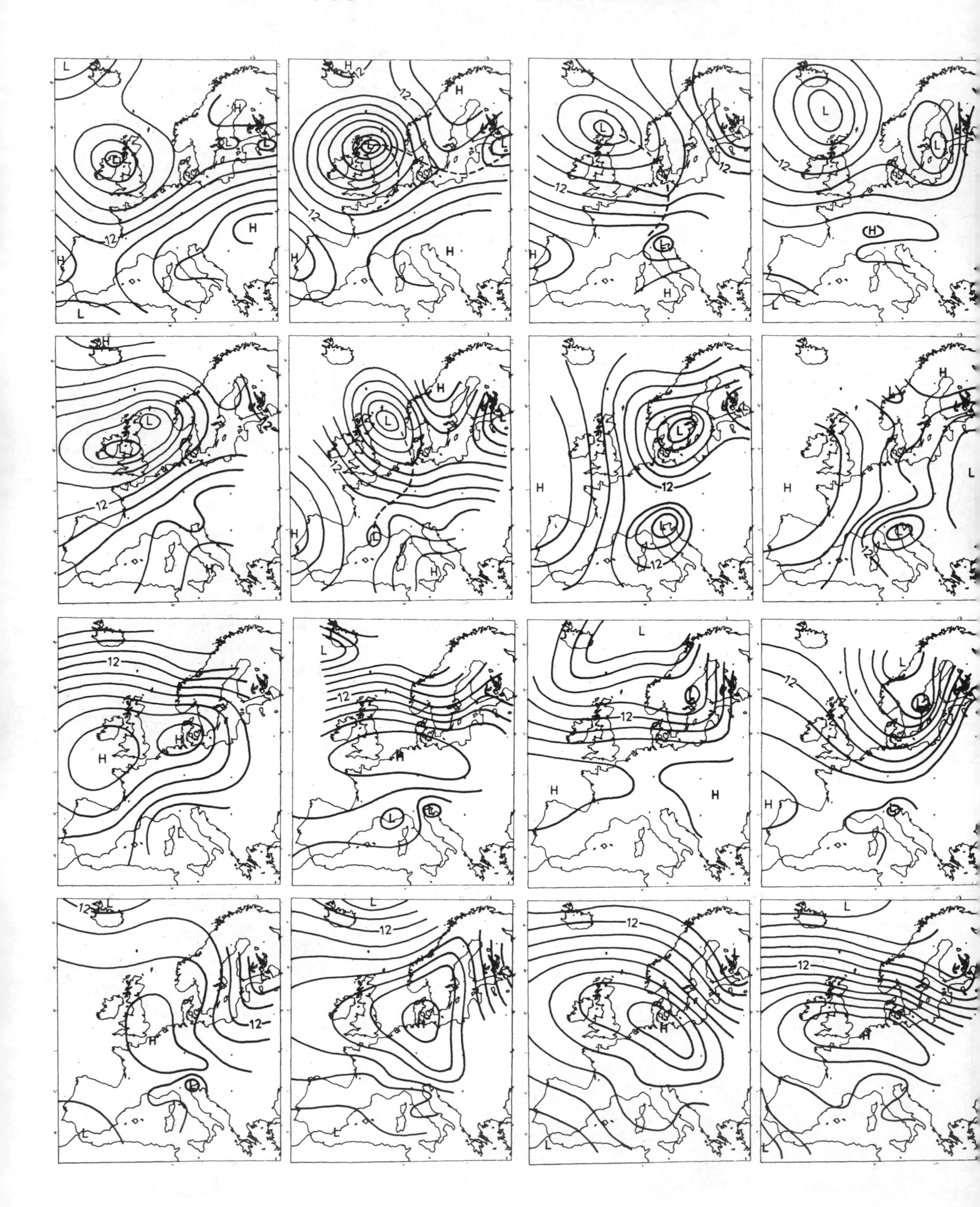

L
H
12

12
L
H

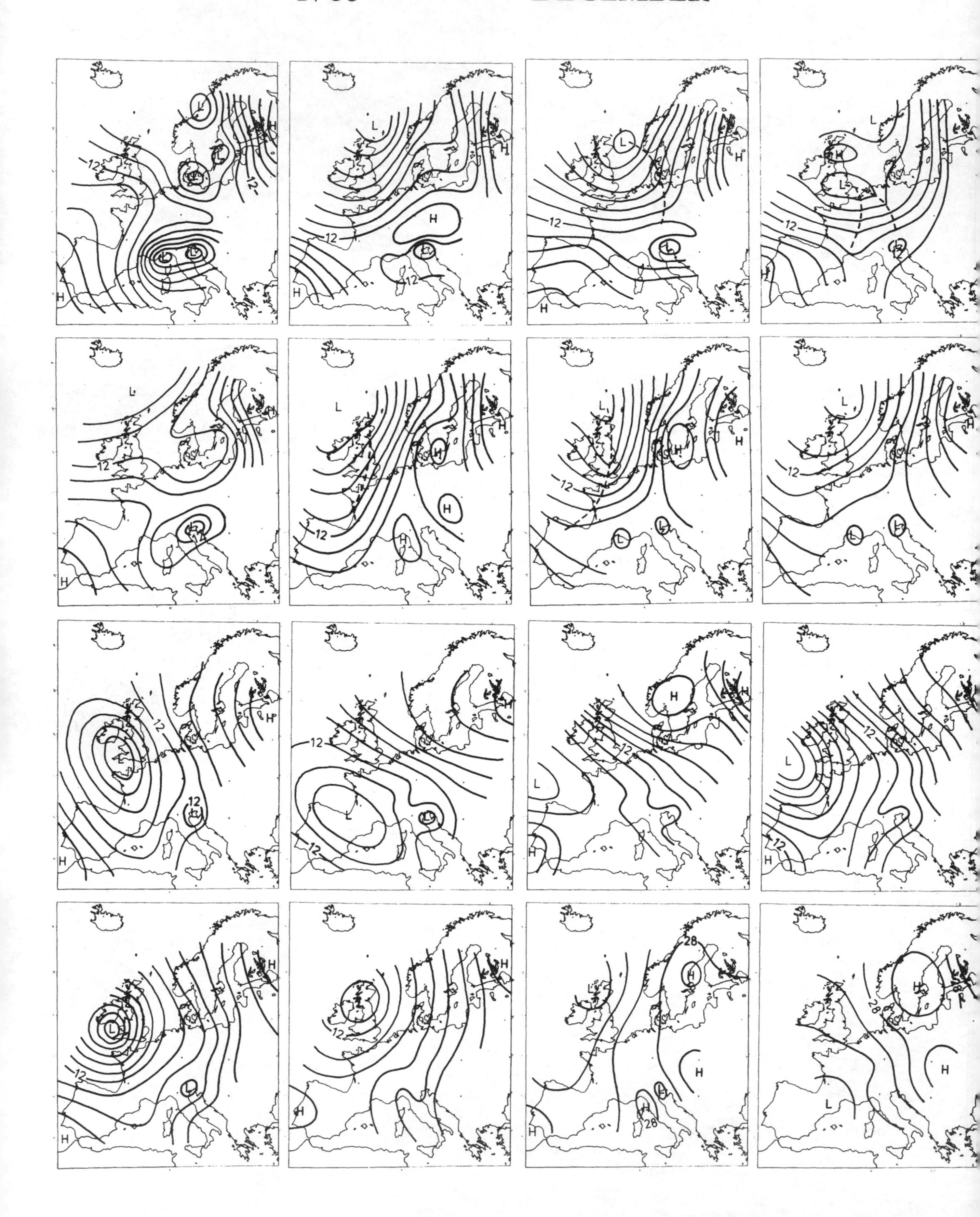

L
H
12

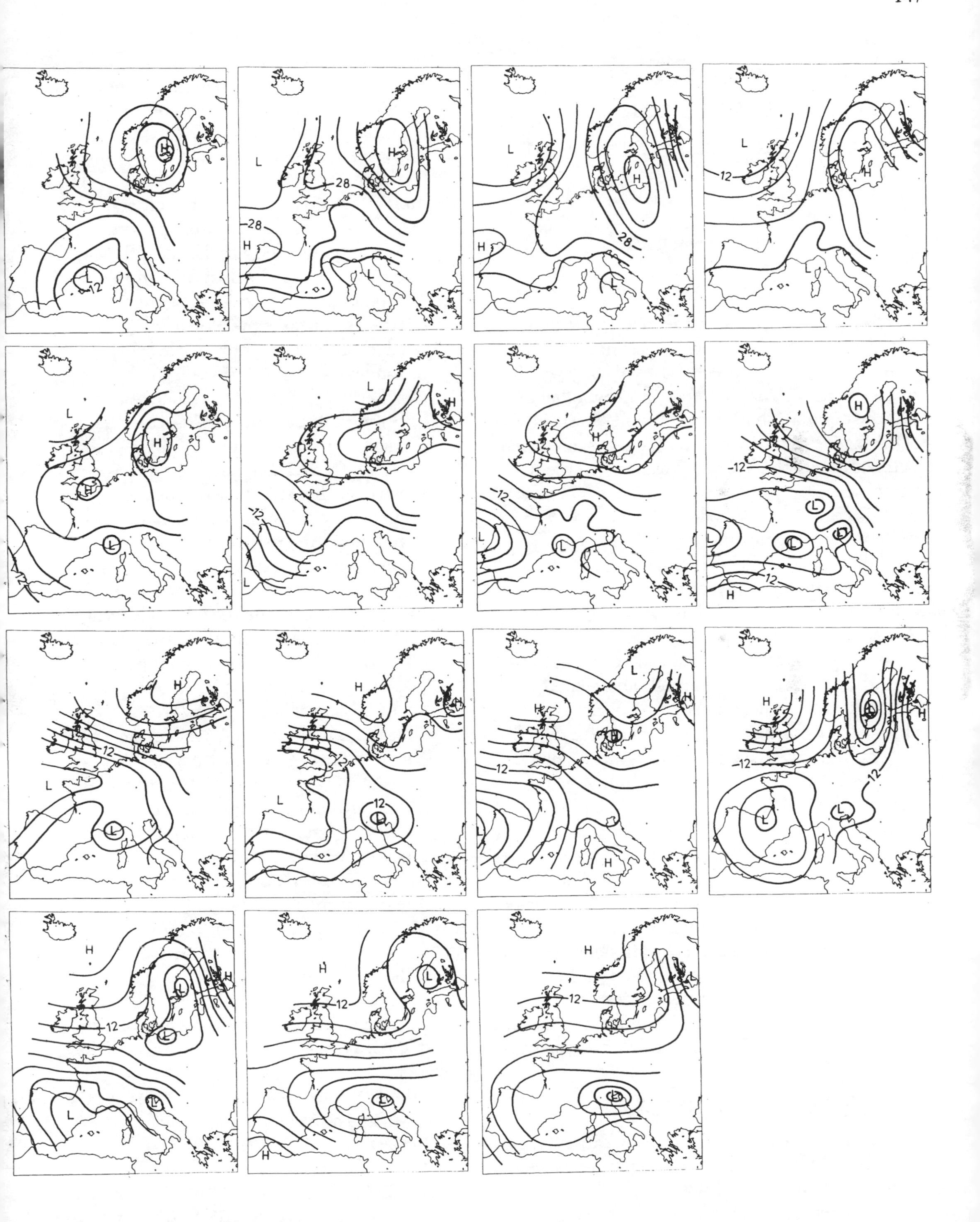

H
L
12
L
H
28
28
H
L
L
H
H
28
L
L
12
H
L
L
H
H
L
L
H
12
L
H
12
H
12
L
L
L
12
L
H
H
12
L
L
12
H
L
H
H
12
L
L
H
12
L
H
H
12
H
L
H
H
12
L
12
L
H
L
12
L
L
L
H
12
L
L
H
12
L

6

Weather types and circulation patterns of the 1780s

Besides providing us with the material to make a detailed history of weather and climate over Europe during the 1780s, a series of daily weather maps can also be used to determine whether or not the circulation patterns of the 1780s have any significance or bearing on the kind of atmospheric behaviour which has been prevalent in recent years.

It is convenient to reduce the mass of meteorological information depicted on a sequence of daily charts into a less unwieldy form by classifying the analysed synoptic situations according to schemes of defined weather types and circulation patterns. This opens up the way for the daily weather situations to be investigated according to statistical methods and techniques which have been developed in synoptic climatology during the twentieth century.

The area covered by the charts allows two well-tested methods of classification to be applied: the Lamb system of British Isles weather types (1972); and the scheme of European large-scale weather patterns (*Grosswetterlagen*) developed by Baur (1947) and later revised and extended by Hess and Brezowsky (1977). By using both classifications the circulation patterns depicted on the charts can be examined from two points of view, i.e. according to the weather conditions over both the British Isles and Central Europe. A more comprehensive picture about the characteristic features of the circulation can thus be obtained.

Lamb British Isles weather types

Dealing first with the British Isles weather types, this classification comprises seven main types, which are defined as follows:

W Westerly: high pressure to the south, sometimes also to the south-west and south-east, and low pressure to the north of the British Isles. Sequences of depressions and ridges travelling eastwards across the North Atlantic and further eastward.

NW North-westerly: Azores anticyclone displaced north-east towards the British Isles or northwards over the Atlantic west of Ireland. Depressions, often forming near Iceland, travel south-east or east-south-east towards the North Sea and reach their greatest intensity over Scandinavia or the Baltic Sea.

N Northerly: high pressure to the west and north-west of the British Isles, particularly over Greenland, and sometimes extending as a continuous belt south over the Atlantic towards the Azores. Low pressure over the Baltic Sea, Scandinavia and the North Sea. Depressions move southwards or south-eastwards from the Norwegian Sea.

E Easterly: anticyclones over Scandinavia; ridges may extend from there to Iceland or northern Britain. Depressions over western North Atlantic and in Azores–Spain–Biscay area.

S Southerly: high pressure over central and northern Europe. Atlantic depressions blocked west of the British Isles or travel northwards or north-eastwards off western coasts. Less persistent than other types and occurs mainly as a variant between Westerly or Easterly types; very rare in summer.

A Anticyclonic: anticyclones centred over, near, or extending over the British Isles. Cols over the region between two anticyclones may be included.

C Cyclonic: depressions stagnating over, or frequently passing across, the British Isles.

Certain hybrids between wind direction types and either the cyclonic or anticyclonic types are also recognised and abbreviated as follows:

SW, NE and SE
CW, CNW, CN, CE, CS, CSW, CNE and CSE
AW, ANW, AN, AE, AS, ASW, ANE and ASE.

Unclassifiable days are denoted by the letter U.

As the British Isles are centrally placed in the mid-latitude westerly wind belt, as well as being located in one of the sectors around the northern hemisphere most frequently affected by blocking of this flow, variations in the circulation over the more extensive eastern North Atlantic-European region are also well registered by this classification. In addition, it appears that the number of westerly wind days recorded over the British Isles during a period such as a year or a decade provides a useful index of the state of the circulation even further afield; in effect, the Lamb classification keeps a finger on the pulse of global climate. Using this classification Lamb has prepared a catalogue of daily circulation types from 1861 to 1971 (Lamb, 1972) which he keeps up-dated in *Climate Monitor*, 1972–86.

For convenience of handling the working copies of the charts for the 1780s and subsequent analysis of the derived statistics, the classification is usually carried out in monthly sets. As an example of the process, the daily synoptic weather situations depicted on the maps for May 1783 were classified as given in Table 6.1. Using a weighted scoring scheme, whereby hybrid days are allocated either in halves or thirds, a frequency table of the seven main types was obtained (see Table 6.2). The complete daily catalogue of the Lamb British Isles weather types for the period 1781 to 1785 is given in Tables 6.3–7.

This kind of classificatory and statistical treatment has resulted in sets of monthly frequencies for the seven main British Isles weather types for the years from 1781 to 1785 (see Tables 6.8–12), from which annual and period mean values can be obtained and compared with long-period averages and extremes (see Table 6.13). These are providing a quantitative basis for comparing the characteristic features of the circulation in the 1780s with those of other periods in the synoptic record from 1861 to the present day.

One of the most interesting results to emerge from this research so far is the extremely low frequency of the westerly weather type over the British Isles, the mean value for the five-year period, 1781–5, being only 66 days per year (see Table 6.13). This represents a substantial decline or blocking of the prevailing westerlies, which in turn produced occasional spells of extreme warmth or cold, wetness or dryness over the region.

This weakening of the zonal or west-east flow over the British Isles region in the 1780s and its resulting effect on the weather and climate is of particular relevance to current studies of climatic change, in view of the fact that a similar trend has been occurring in recent years. Since 1955 the frequency in the number of westerly-wind days over the British Isles has been steadily decreasing, averaging 80 days per year in the 1960s, 73 days in the 1970s and 67 days in the 1980s. In contrast, the circulation over the region during the first half of the twentieth century was notable for its strong zonal flow, with several decades averaging over 100 westerly-wind days per year.

Table 6.1 *Lamb British Isles weather types for May 1783.*

Date	Type	Date	Type	Date	Type
1	A	11	W	21	E
2	NE	12	CW	22	CNE
3	A	13	CW	23	NE
4	A	14	W	24	N
5	A	15	W	25	U
6	N	16	W	26	CE
7	A	17	NW	27	CNE
8	A	18	AN	28	NE
9	SW	19	ANE	29	CNE
10	W	20	AE	30	C
				31	A

Table 6.2 *Example of weighted scoring system used with the Lamb British Isles weather types for May 1783.*

											Total
N	½	1	1	½	½	1	1	1			6½
NW	1										1
N	½	1	½	⅓	⅓	½	1	⅓	½	⅓	5⅓
E	½	⅓	½	1	⅓	½	½	⅓	½	⅓	4⅚
S	½										½
A	1	1	1	1	1	1	½	⅓	½	1	8⅓
C	½	½	⅓	½	⅓	⅓	1				3½
U	1										1
											31

Table 6.3 *Catalogue of Lamb circulation types over the British Isles for 1781.*

	Jan	Feb	Mar	Apr	May	Jun	Jul	Aug	Sep	Oct	Nov	Dec
1	W	W	W	S	E	ASW	S	W	W	W	W	A
2	CW	AW	AW	A	E	S	C	AW	CSW	W	W	ASE
3	A	AW	W	A	E	CS	CW	A	SW	A	NW	SE
4	A	W	W	A	E	C	W	A	C	A	A	C
5	AW	C	AW	CSE	A	C	ASW	A	C	A	C	C
6	AW	C	ANW	C	ANE	C	S	A	C	A	C	U
7	AW	SW	ANW	C	AE	C	C	E	W	A	CN	SE
8	A	SW	AW	CW	AE	C	C	C	AW	A	NW	E
9	A	SW	AW	SW	E	CE	C	C	A	A	AW	E
10	A	W	A	C	AE	CE	W	C	A	ASW	CW	SE
11	AE	W	A	C	CE	C	W	C	A	ASW	CNW	SE
12	AE	W	A	N	CSE	CE	W	C	A	NW	NW	SE
13	E	W	A	A	C	C	W	C	AE	AW	NW	SE
14	E	W	A	W	C	C	A	CW	AS	ASW	CW	CS
15	E	W	A	W	CE	C	ANW	C	S	W	CNW	S
16	CE	NW	ASE	SW	CE	C	A	C	C	ANW	NW	CSW
17	CE	NW	A	U	E	CSW	A	C	CW	A	C	SW
18	C	NW	A	S	CE	C	A	U	NW	ANW	N	S
19	N	N	AW	W	C	CSE	A	NE	W	NW	A	S
20	C	A	ASW	W	E	CE	A	N	W	NW	W	SW
21	U	A	AW	W	A	E	A	NW	NW	N	W	SW
22	E	N	A	W	AE	U	A	C	NW	N	NW	SW
23	E	ANW	A	AW	AE	C	A	C	NW	AN	CS	SW
24	C	CNW	A	A	ASE	E	AW	C	NW	A	SW	SW
25	C	NW	A	ANW	A	C	W	C	N	A	AW	W
26	N	NW	NE	A	A	AE	AW	C	N	A	SW	C
27	W	NW	NE	A	A	A	W	C	N	A	C	W
28	W	NW	ANE	A	AS	ANW	W	C	N	C	U	W
29	W		NE	E	AS	AW	W	C	W	CN	A	CW
30	W		NE	E	ASE	AW	ASW	U	CW	N	S	C
31	W		ASE		ASE		SW	C		NW		C

Further evidence of the relevance of this research to current studies is given by an examination of all five-year periods in the now extended British Isles synoptic record, i.e., 1781–5; 1861–1986. This analysis shows that the two five-year periods having the lowest mean frequency of the westerly type occur over a time span of nearly 200 years. The most recent one, 1976–80, with a value of 66.7 days per year, is almost as low as that of 1781–5, with 66.2 days per year. A comparison of the frequencies of the other British Isles weather types in these two periods shows that there were further similarities in the circulation patterns that accompanied the declines in the westerly type two centuries ago and today. Both the five-year periods 1781–5 and 1976–80 include a year with an extreme minimum value of the westerly type, namely, 1785 with 45 days, the absolute minimum, and 1980 with 52 days, the lowest since the official record began in 1861. For purposes of comparison the means of both periods have been related to the averages for 1900 to 1954, a climatic period when the westerly weather type was very prevalent and which for some time was regarded as the 'normal' mode of circulation over the British Isles region (see Table 6.14.) It can be seen from this table that in both five-year periods under review, there was a decrease of one-third in the westerly type, with mean annual frequencies of only 66 or 67 days, compared with 99 days per year during the so-called 'normal' period. These notable decreases in zonal circulation over the British Isles region both in the 1780s and recent years were compensated for by increases in blocked or stationary weather patterns, which resulted in changes in the frequencies of airstreams from other directions. Almost equal increases occurred in the

Table 6.4 *Catalogue of Lamb circulation types over the British Isles for 1782.*

	Jan	Feb	Mar	Apr	May	Jun	Jul	Aug	Sep	Oct	Nov	Dec
1	SW	A	W	C	E	U	W	C	AW	NW	N	A
2	W	W	W	CN	NE	ANW	NW	C	A	C	C	A
3	W	C	W	CNW	E	ANW	NW	C	AS	N	C	A
4	W	C	AW	C	E	ANW	NW	C	S	A	CN	A
5	W	E	W	C	E	ANW	C	CNW	S	U	C	SE
6	W	NE	W	C	NE	A	E	C	A	ANE	N	S
7	W	ANE	W	NE	A	ASW	CE	C	SE	NE	AN	SE
8	W	E	CW	AE	SE	S	U	CN	ASE	A	A	A
9	CW	E	N	U	CE	C	W	N	ASE	S	A	U
10	A	E	C	N	C	C	NW	NW	AE	CSE	NE	A
11	AW	E	C	C	C	C	C	CNW	AE	E	U	A
12	AW	E	U	CE	C	C	C	C	AE	NE	SW	SW
13	AW	NE	N	CN	W	C	C	C	ANE	A	A	W
14	ASW	AE	NE	CN	C	C	C	C	E	A	AW	NW
15	AW	A	A	E	C	CW	A	C	SE	AW	W	CNW
16	W	A	AN	E	C	ASE	ASE	C	SE	AW	AN	W
17	NW	A	AN	E	C	SE	U	CNW	C	A	AN	W
18	W	A	NW	E	C	C	A	NW	C	W	N	W
19	W	A	N	E	N	AW	A	W	CN	NW	N	W
20	W	A	N	E	C	A	A	W	C	NW	A	W
21	W	SW	CE	CS	C	A	A	SW	C	NW	A	AW
22	W	SW	CE	C	C	A	S	CW	C	W	SW	A
23	W	C	NE	CSE	C	A	CSW	W	C	NW	S	W
24	W	W	CN	CSE	C	ASE	CSW	CSW	CSW	NW	C	NW
25	CW	W	CN	E	W	S	W	W	SW	W	C	ANW
26	NW	W	SW	E	W	W	U	W	CSW	AW	A	AW
27	C	CW	SW	E	SW	ANW	C	W	CW	AW	S	AW
28	C	AW	W	E	C	AW	C	N	CW	CW	C	ANW
29	C		W	E	C	W	N	C	W	CN	C	ANW
30	NE		W	E	C	W	NW	NW	N	C	U	N
31	ANE		C		N		CNW	NW		C		AN

frequencies of the north-westerly and easterly types. The cyclonic type was also more frequent in both periods, especially in 1781–5 when three out of the five years analysed were very cyclonic. Although the northerly type was more frequent in 1781–5, it was very near to the 'normal' value in 1976–80. The southerly type presented an interesting reversal, being somewhat less frequent in 1781–5 but decidely more frequent in 1976–80. The frequencies of the anticyclonic type were also reversed but in the opposite sense and not to such a marked degree.

PSCM *indices*

The classification of Lamb British Isles weather types allows further synoptic-climatological studies to be made. For example, for some purposes it is desirable to have a ready indication of the general character of a month, season, year or period of years. The indices of progression, meridionality and cyclonicity or *PSCM* indices devised by Murray and Lewis (1966) fulfil this function.

The *P* index is a measure of the frequency difference between days of progressive (that is westerly) and blocked (that is easterly or meridional) circulation types, with *P* being positive when progressive types predominate over blocked types. The *S* index is a measure of the frequency difference between days of southerly and northerly types, with *S* being positive when the bias is towards southerliness. The *M* index is a measure of *total* meridionality, that is, the total frequency of both southerly and northerly types; and the *C* index is a measure of the frequency difference between days of cyclonic and anticyclonic types, with *C* being positive when the bias is towards cyclonicity.

Table 6.5 *Catalogue of Lamb circulation types over the British Isles for 1783.*

	Jan	Feb	Mar	Apr	May	Jun	Jul	Aug	Sep	Oct	Nov	Dec
1	A	W	C	AW	A	ASE	A	A	A	A	C	SW
2	A	W	NW	A	NE	SE	CS	CS	ASW	A	CS	S
3	AS	W	C	A	A	CSE	SW	C	W	W	SE	S
4	SW	SW	U	A	A	C	AW	C	W	W	A	S
5	W	W	C	A	A	C	AW	S	NW	CW	SE	S
6	W	W	C	A	N	W	AW	C	C	W	SE	A
7	NW	C	C	A	A	AW	AW	C	NW	NW	SE	A
8	CW	C	C	ASW	A	A	AS	CW	W	NW	SE	A
9	CW	C	CW	ASW	SW	A	S	W	CNW	W	A	A
10	CW	C	N	S	W	AW	CS	CW	NW	W	A	A
11	CW	SW	NE	W	W	C	CS	N	U	A	N	A
12	C	C	E	U	CW	C	U	AN	C	SE	N	A
13	CW	C	ANE	A	CW	C	SE	A	W	S	C	A
14	CW	E	AE	AW	W	C	SE	A	W	A	C	A
15	C	A	AE	W	W	C	CSE	AW	W	A	N	A
16	N	A	A	ANW	W	C	C	AW	U	A	C	A
17	NE	A	A	AW	NW	C	U	ASW	A	A	C	A
18	CN	A	A	A	AN	C	ASW	AS	U	AS	W	A
19	CN	ANW	A	SW	ANE	C	W	S	CS	SW	C	A
20	CE	W	A	W	AE	N	SW	AS	C	SW	C	A
21	C	W	S	AN	E	A	CSW	C	CSW	S	CN	ASW
22	C	W	U	AN	CNE	A	SW	W	C	W	A	C
23	CNW	C	A	NE	NE	ASW	NW	W	C	W	AN	C
24	C	CN	ANW	NE	N	A	W	C	AW	W	AN	C
25	C	N	C	E	U	AW	W	C	ASW	W	A	CE
26	CSW	ANW	C	AE	CE	A	S	C	A	W	A	CE
27	C	U	N	E	CNE	A	C	C	A	CW	A	E
28	CNW	C	ANW	SE	NE	AW	C	CE	A	W	S	E
29	W		A	SE	CNE	A	C	C	A	W	SW	U
30	W		W	ANE	C	A	W	C	AE	W	ASW	A
31	W		W		A		W	U		C		SE

The *P* index curve of five-year running means in Figure 6.1 shows a decrease of progression from 1861 to the late 1880s, then a rise to the overall maximum of progressiveness in the early 1920s, followed by a decrease, apart from a temporary increase around the early 1950s, to the low values of recent years. It is striking that the period mean value of 1781–5 of −36 (the dashed line), the second lowest on record, equals that of 1981–5 exactly 200 years later; the overall extreme being −41 in 1968–72. These values indicate a very pronounced dominance of blocked or non-progressive weather types during the 1780s and in recent years.

Although generally depressed in magnitude, the *P* index curve in Figure 6.2, showing seasonal variations of progression for 1781–5, suggests that even during a period of reduced westerly type a fundamental pattern continues to exert control over much of the year. However, there are several interesting points which stand out when this curve is compared with the long-period mean curve. For example:

(i) maximum of blocking occurring in *early* rather than late spring;

(ii) a sharp increase in progression occurring in early summer, with the July value for 1781–5 comparable to the long-period mean value, indicating that the timing and magnitude of the onset of the so-called 'European monsoon' was similar to average in 1781–5;

(iii) a marked departure from the mean curve in December, showing that there was a pronounced secondary maximum of blocking during the early winter period, indicating an early start to cold-season continental conditions in 1781–5. This is an interesting

Table 6.6 *Catalogue of Lamb circulation types over the British Isles for 1784.*

	Jan	Feb	Mar	Apr	May	Jun	Jul	Aug	Sep	Oct	Nov	Dec
1	CSE	ANW	A	N	AN	U	ANW	W	U	A	C	AN
2	CS	N	A	C	AN	A	AN	A	AW	A	NW	A
3	C	A	SW	CE	AN	AE	A	A	SW	S	SW	NW
4	U	AW	CSW	A	A	E	A	A	SW	SE	CSE	NW
5	A	N	CSW	AW	ANW	E	A	W	AS	CSE	E	C
6	A	N	S	A	A	U	A	W	ASW	SE	E	C
7	S	N	CSE	A	ASW	C	A	NW	A	AE	AE	C
8	A	N	C	A	W	C	NE	ANW	AS	E	A	CNE
9	A	CN	C	N	W	C	ANW	A	AW	E	C	N
10	A	N	N	N	W	C	A	C	AW	E	C	N
11	A	CN	AN	CNW	ANW	NW	AW	W	A	A	C	NE
12	A	CNE	NW	CN	AW	C	W	AW	ASW	A	N	CN
13	AW	E	ANE	CN	AW	C	AW	AW	U	A	NW	NE
14	W	E	A	C	W	W	ANW	AW	A	A	W	NE
15	W	E	A	C	AW	W	A	A	A	A	W	ANE
16	C	E	AE	E	A	W	ANW	NW	A	A	W	NE
17	CN	E	AE	C	A	W	ANW	N	A	A	CW	A
18	CN	A	E	NW	A	W	W	A	SW	A	C	E
19	N	A	E	C	A	SW	C	A	S	A	N	AN
20	N	S	CN	W	AW	W	NW	AE	C	NW	A	A
21	C	S	C	NW	AW	C	W	A	W	AW	A	E
22	NE	SW	E	C	AW	C	NW	E	C	NW	AW	NE
23	N	W	S	NW	W	NW	NW	N	N	CNW	W	AN
24	N	SW	CSE	NW	AW	CN	W	C	SW	N	W	A
25	N	W	CE	W	CW	NW	W	CNE	CSW	N	CW	A
26	NE	NW	E	W	C	C	CSW	N	C	A	AW	A
27	NE	N	NE	CW	W	N	W	A	C	A	ASW	A
28	E	ANE	AE	N	W	N	SW	CW	U	A	W	AE
29	E	A	AE	A	SW	NW	C	W	E	A	ANE	E
30	AN		E	A	SW	AN	C	A	AE	A	U	C
31	ANW		ANE		C		CNW	CS		C		E

and certainly genuine characteristic of the Little Ice Age period.

Grosswetterlagen

The Lamb British Isles weather type classification was particularly well-suited for establishing a long series of circulation patterns from 1861, since the daily weather maps that were available in the earlier part of the official record covered little more than the British Isles themselves. However, the charts which are now being constructed for the 1780s have a synoptic observational coverage which allows the analysis to be made over a much wider area of Europe. Accordingly, a broader-based approach to the classification process is now possible by applying *Grosswetterlage* principles to the synoptic weather patterns which are emerging from the 1780s. The first years of an expanding catalogue of *Grosswetterlagen* covering Europe from 1781 are given in Tables 6.15–19. This may be compared to the main catalogue of *Grosswetterlagen* from 1881 to 1986 published by the Deutscher Wetterdienst (Hess and Brezowsky, 1977, and *Die Grosswetterlagen Mitteleuropas*, 1977–86).

In this classification the term *Grosswetterlage* refers to a large-scale weather pattern in which the positions of the primary steering centres, highs and lows, together with the main features of the circulation, are considered to remain essentially the same for several days.

The patterns are defined as follows, with the letters A or Z being appended to directional types according to whether the curvature of the isobars over Central Europe is anticyclonic or cyclonic.

Table 6.7 *Catalogue of Lamb circulation types over the British Isles for 1785.*

	Jan	Feb	Mar	Apr	May	Jun	Jul	Aug	Sep	Oct	Nov	Dec
1	E	A	A	N	AE	CNW	C	U	C	A	C	NW
2	CE	A	A	NE	ANE	NW	C	AE	C	A	C	W
3	CE	U	A	A	A	C	NW	CE	SW	C	CW	CNW
4	C	NW	AE	AW	AW	C	CNW	C	S	SW	CW	C
5	C	CE	AE	A	U	C	NW	C	CSW	CSW	C	U
6	N	C	A	A	AW	U	NW	C	C	CSW	CNW	SW
7	AW	CN	AE	A	AW	C	N	W	SW	SW	N	CW
8	SW	N	AE	A	AW	C	ANW	CW	U	W	AN	W
9	SW	NW	E	A	A	W	ANW	N	S	W	A	C
10	A	C	A	A	SE	AW	NW	NW	C	W	A	E
11	A	SE	A	A	SE	AW	AN	W	CS	W	AW	E
12	A	A	NE	AW	ASE	A	U	C	SW	SW	NW	SE
13	SE	A	NE	A	A	A	U	C	SW	W	A	C
14	SE	ANE	AN	A	AW	AS	C	N	CS	W	A	C
15	E	N	ANE	AW	ANW	U	C	NW	C	AW	A	CS
16	S	N	ANE	A	NW	NW	C	C	SW	A	A	SE
17	SW	CN	A	AS	CN	A	C	C	C	A	A	E
18	S	CNE	A	CSW	NW	A	C	N	C	AE	U	A
19	S	C	ANW	CW	NW	A	CW	N	SW	A	C	SW
20	S	CN	AE	AW	N	ANW	C	N	C	AE	CW	SW
21	S	C	N	AW	NW	A	C	NE	C	A	CN	ASW
22	SW	C	ANE	ANW	AW	A	CN	N	U	AW	A	A
23	SW	U	ANE	NW	AS	A	NW	U	C	A	A	A
24	SW	CW	N	A	S	A	CW	C	C	W	A	E
25	SW	A	A	A	W	A	C	C	C	NW	W	CE
26	AS	SE	N	A	AW	A	C	N	NW	NW	W	CE
27	W	SE	NE	A	C	A	C	ANW	NW	NW	NW	AE
28	W	SE	N	A	C	ASE	C	A	AN	W	C	ANE
29	NW		NE	A	N	SE	C	S	A	NW	C	E
30	C		N	ANE	AN	A	C	CSW	U	W	N	E
31	N		N		NW		C	W		C		E

HM — Anticyclone over Central Europe
HNA and HNZ — Anticyclone over the Norwegian Sea
HFA and HFZ — Anticyclone over Scandinavia
HNFA and HNFZ — Anticyclone over the Norwegian Sea and Scandinavia
HB — Anticyclone over the British Isles
BM — Ridge of high pressure over Central Europe
TM — Depression over Central Europe
TB — Depression over the British Isles
TRM — Trough of low pressure over Central Europe
TRW — Trough of low pressure over Western Europe

SWA and SWZ — South-westerly
SA and SZ — Southerly
SEA and SEZ — South-easterly
NEA and NEZ — North-easterly
NA and NZ — Northerly
NWA and NWZ — North-westerly
WA and WZ — Westerly
WW — Blocked westerly with high pressure over Russia and a meridionally orientated Polar Front over Europe and the North Sea
WS — Westerly with the sub-tropical high and Polar Front south of their normal positions
U — Unclassifiable

For general statistical studies of atmospheric circulation over Europe, related *Grosswetterlagen*

Table 6.8 *Lamb British Isles weather types: monthly and annual frequencies for 1781.*

	W	NW	N	E	S	A	C	U
January	8	0	2	7	0	7½	5½	1
February	10½	8	2	0	1½	3½	2½	0
March	6⅓	1	2⅓	3	1	17⅓	0	0
April	8	½	1	2⅓	3⅓	9	4⅚	1
May	0	0	⅓	13⅙	2⅓	9⅚	5⅓	0
June	1⅔	½	0	4⅚	2½	3⅓	16⅙	1
July	12⅔	½	0	0	3⅙	10⅙	4½	0
August	2	1	1½	1½	0	4½	18½	2
September	7⅓	5	4	½	2⅓	5½	5⅓	0
October	4½	5	4	0	1	15	1½	0
November	7	6½	2	0	2½	4	7	1
December	6⅚	0	0	5⅓	10⅙	1⅓	6⅓	1
Year	74⅚	28	19⅙	37⅔	29⅚	91	77½	7

can be grouped into three main circulation patterns:

(i) Zonal, comprising all westerly patterns, i.e. WA, WZ, WS and WW;

(ii) Mixed, comprising patterns in which the sub-tropical anticyclone has shifted northwards to about 50 °N, but is not blocking, i.e. NWA, NWZ, HM, BM, SWA, SWZ, and TM;

(iii) Meridional, comprising all northerly, north-easterly, easterly, south-easterly and southerly situations with blocking highs lying between 50 ° and 65 °N over the eastern North Atlantic-European sector, i.e. NA, NZ, HNA, HNZ, HB, TRM, NEA, NEZ, HFA, HFZ, HNFA, HNFZ, SEA, SEZ, SA, SZ, TB, and TRW.

Annual and period mean frequencies of these three main circulation patterns for 1781–5 are given in Table 6.20 together with the long-term mean values for the 100-year period, 1881–1980. Examination of this table shows that in the five-year period, 1781–5, the circulation over Europe fluctuated widely from year to year. For instance, although the period mean frequency of zonal circulation patterns was only about 5 per cent less than average, the annual value in 1781, 55 days, the second lowest on record (cf. 51 days in 1901), was followed by a value of 123 days in 1782 which indicates that zonal-type flow over Europe that year was over 25 per cent more frequent than average. Comparable fluctuations occurred in the mixed circulation patterns. For example, whilst annual values in 1781, 1783 and 1784 were 25 to 30 per cent more frequent than average, those in 1782 and 1785 were about 20 per cent less frequent. With the meridional circulation patterns, annual values in 1781, 1782 and 1785 were only between 5 and 10 per cent different from average, but those in 1783 and 1784

Figure 6.1. *P* index: graph of five-year running means, 1861–1978 (from Kington, 1980).

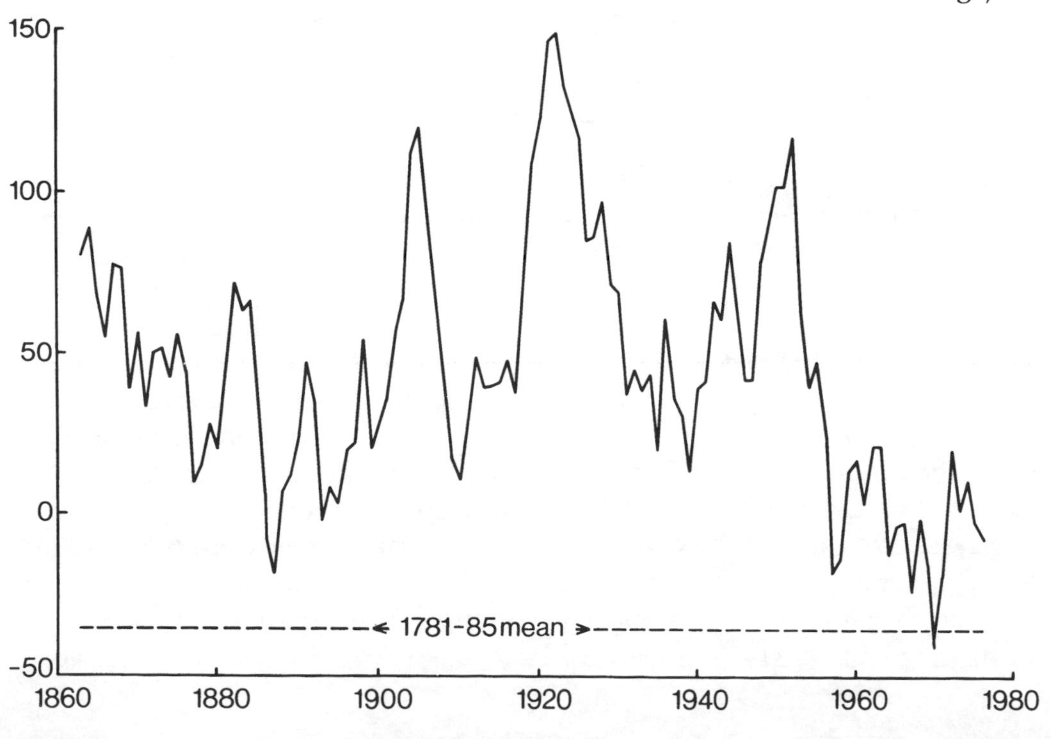

Table 6.9 *Lamb British Isles weather types: monthly and annual frequencies for 1782.*

	W	NW	N	E	S	A	C	U
January	18⅚	2	⅚	⅚	⅚	3⅔	4	0
February	6	0	1⅓	7⅚	1	8⅓	3½	0
March	11	1	7	2	1	2½	5½	1
April	0	½	3	14⅙	1⅙	½	9⅔	1
May	3½	0	3	6	1	1	16½	0
June	4⅚	2½	0	1⅙	3½	9½	7½	1
July	3⅔	5½	1	1⅚	2	5⅓	8⅔	3
August	7⅓	5½	2½	0	⅚	0	14⅚	0
September	3⅔	0	1⅚	5	5⅚	5	8⅙	0
October	5½	6	2⅚	2⅔	1⅓	7⅓	4⅓	1
November	2½	0	6½	½	3	8	7½	2
December	9	4	1½	1	2½	11½	½	1
Year	75⅚	27	31⅓	43	24	63⅙	90⅔	10

Table 6.10 *Lamb British Isles weather types: monthly and annual frequencies for 1783.*

	W	NW	N	E	S	A	C	U
January	8⅚	2	2½	1	1⅓	2½	12⅚	0
February	9	1	1½	1	1	5	8½	1
March	2½	2	2⅚	2⅚	1	9⅓	8½	2
April	5⅔	½	2⅓	4⅚	3⅙	12½	0	1
May	6½	1	5⅓	4⅚	½	8⅓	3½	1
June	3⅓	0	1⅙	1	1½	11⅔	11⅓	0
July	9⅙	1	0	1⅓	7½	3⅚	6⅙	2
August	5⅓	0	1½	½	3⅚	5⅚	13	1
September	7½	3½	0	½	1½	7⅔	6⅓	3
October	15	2	0	½	4	7½	2	0
November	1⅚	0	4½	2½	4⅚	8⅓	8	0
December	⅚	0	0	3½	5⅓	16⅓	4	1
Year	75½	13	21½	24½	35½	98⅚	84⅙	12

Table 6.11 *Lamb British Isles weather types: monthly and annual frequencies for 1784.*

	W	NW	N	E	S	A	C	U
January	2½	½	8	3⅚	1⅚	8½	4⅚	1
February	3½	1½	8⅔	5⅔	3	5⅓	1⅓	0
March	1⅙	1	3⅙	9⅓	3⅚	7⅙	5⅓	0
April	4	4½	5	1½	0	6½	8½	0
May	12⅓	1	1½	0	1⅓	12⅓	2½	0
June	6½	4	3	2½	½	2	9½	2
July	7⅚	6	1	½	⅚	11	3⅚	0
August	7	2½	3⅓	1⅚	½	12½	3⅓	0
September	5½	0	1	1½	5	9⅔	4⅓	3
October	½	2½	2	4⅚	2⅓	17	1⅚	0
November	8⅚	2	2⅓	3⅙	1⅙	5⅙	6⅓	1
December	0	2	7⅙	7⅔	0	9⅓	4⅚	0
Year	59⅔	27½	46⅙	42⅓	20⅓	106½	56½	7

Table 6.12 *Lamb British Isles weather types: monthly and annual frequencies for 1785.*

	W	NW	N	E	S	A	C	U
January	6	1	2	4	10	4	4	0
February	½	2	5⅙	3⅙	2	5⅓	7⅚	2
March	0	½	9⅚	6⅚	0	13⅚	0	0
April	3⅓	1½	1⅚	⅚	⅚	20⅚	⅚	0
May	4½	5½	3⅓	2⅙	2⅚	9⅙	2½	1
June	2	3	0	⅚	1⅓	15⅓	5½	2
July	1	6½	2	0	0	1½	18	2
August	3⅚	2½	7½	1½	1⅓	2	10⅓	2
September	3⅓	2	½	0	6⅓	1½	13⅓	3
October	12⅙	4	0	1	2⅙	9	2⅔	0
November	4	2½	3	0	0	11	8½	1
December	4⅓	1½	⅓	9⅚	3⅓	4⅙	6½	1
Year	45	32½	35½	30⅙	30⅙	97⅔	80	14

were over 20 per cent less frequent, with that for the latter year, 107 days, being the fourth lowest on record (cf. 94 days in 1948, 96 in 1943 and 106 in 1883).

In a study on the climatology of blocking action, Brezowsky, Flohn and Hess (1951) showed that during the 70-year period 1881–1950 a periodicity of 22 to 23 years was apparent in the frequency of blocking highs over the eastern North Atlantic-European sector (A+E), with maxima of overlapping 10-year means in 1892, 1915 and 1937 being closely situated to the sunspot maxima in 1893, 1917 and 1937. A similar study has been made of blocking action during the period 1781–5 and the annual frequencies of blocking highs are plotted against yearly sunspot numbers in Figure 6.3. With reference to the finding of Brezowsky *et al.* it is interesting to see how the two sets of data keep in step, with the very high blocking value of 123 days in 1781 closely following the sunspot maximum in 1778, and the low blocking value of 52 days in 1784 corresponding to the sunspot minimum in the same year. As further daily weather maps become available for *Grosswetterlage* classification from 1786 onwards it will become possible to discover more about the nature and validity of this apparent sun–weather relationship.

Table 6.13 *Lamb British Isles weather types: annual and period average frequencies, 1781–5; long-period average frequencies and extremes, 1861–1980.*

	Number of days per year						
	W	NW	N	E	S	A	C
1781–5							
1781	74.8*	28.0	19.2	37.7	29.8	91.0	77.5
1782	75.8	27.0	31.3	43.0	24.0	63.2	90.7
1783	75.5	13.0	21.5	24.5	35.5	98.8	84.2
1784	59.7	27.5	46.2	42.3	20.3	106.5	56.5
1785	45.0	32.5	35.5	30.2	30.2	97.7	80.0
Average	66.2	25.6	30.7	35.5	28.0	91.4	77.8
1861–1980							
Average	91.1	18.3	27.5	28.1	31.0	91.0	64.2
Highest value	128.7 (1923)	42.5 (1973)	48.8 (1919)	57.8 (1963)	54.2 (1924)	129.2 (1971)	100.0 (1872)
Lowest value	51.8 (1980)	5.0 (1911, 1960)	10.2 (1920)	9.7 (1967)	15.5 (1896)	38.2 (1872)	41.7 (1921)

*Decimal values arise because of allocating hybrids of two or three types to the seven basic weather types: Westerly (W), North-westerly (NW), Northerly (N), Easterly (E), Southerly (S), Anticyclonic (A) and Cyclonic (C).

Figure 6.2. *P* index: graphs of the seasonal pattern of progression for the periods 1861–1978 and 1781–5 (from Kington, 1980).

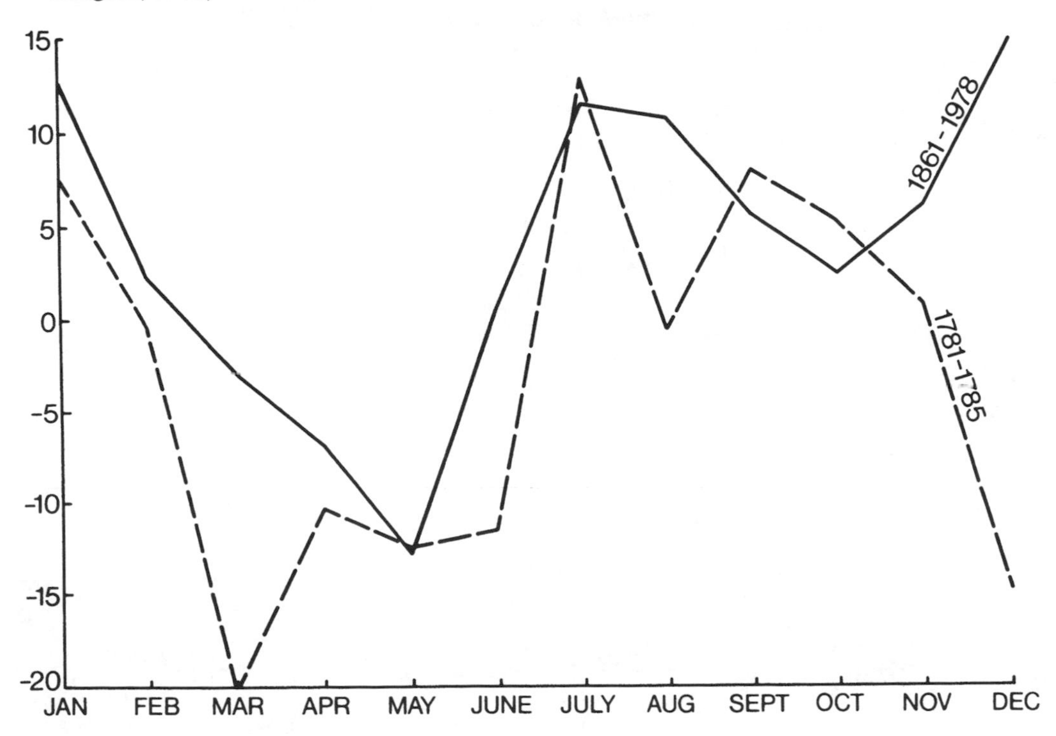

Table 6.14 *Lamb British Isles weather types: period average frequencies for 1900–54; 1781–5; and 1976–80.*

	Number of days per year						
	W	NW	N	E	S	A	C
1900–54	98.5	16.1	26.7	25.2	30.7	89.1	63.9
1781–5	66.2	25.6	30.7	35.5	28.0	91.4	77.8
	(−33)	(+59)	(+15)	(+41)	(−9)	(+3)	(+22)
1976–80	66.7	26.0	27.2	34.8	37.9	83.0	71.3
	(−32)	(+62)	(+2)	(+38)	(+24)	(−7)	(+12)

N.b. The values in brackets are percentage differences from the 1900–54 values.

Future developments

The series of daily historical weather maps from 1781, of which the first five years have been presented and discussed in this book, is allowing a detailed synoptic examination to be made of the weather patterns which occurred over Europe during the decade leading up to the outbreak of the French Revolution (Kington, 1980). The author is continuing with the preparation, analysis and classification of charts for the remaining five years from 1786 to 1790.

This programme of daily weather mapping is also providing the means of making meteorologically well-documented historical case studies of varied events that occurred during the period which were directly or indirectly affected

Figure 6.3. *Grosswetterlagen:* graphs of blocking highs (A+E) and annual mean sunspot numbers, 1780–6.

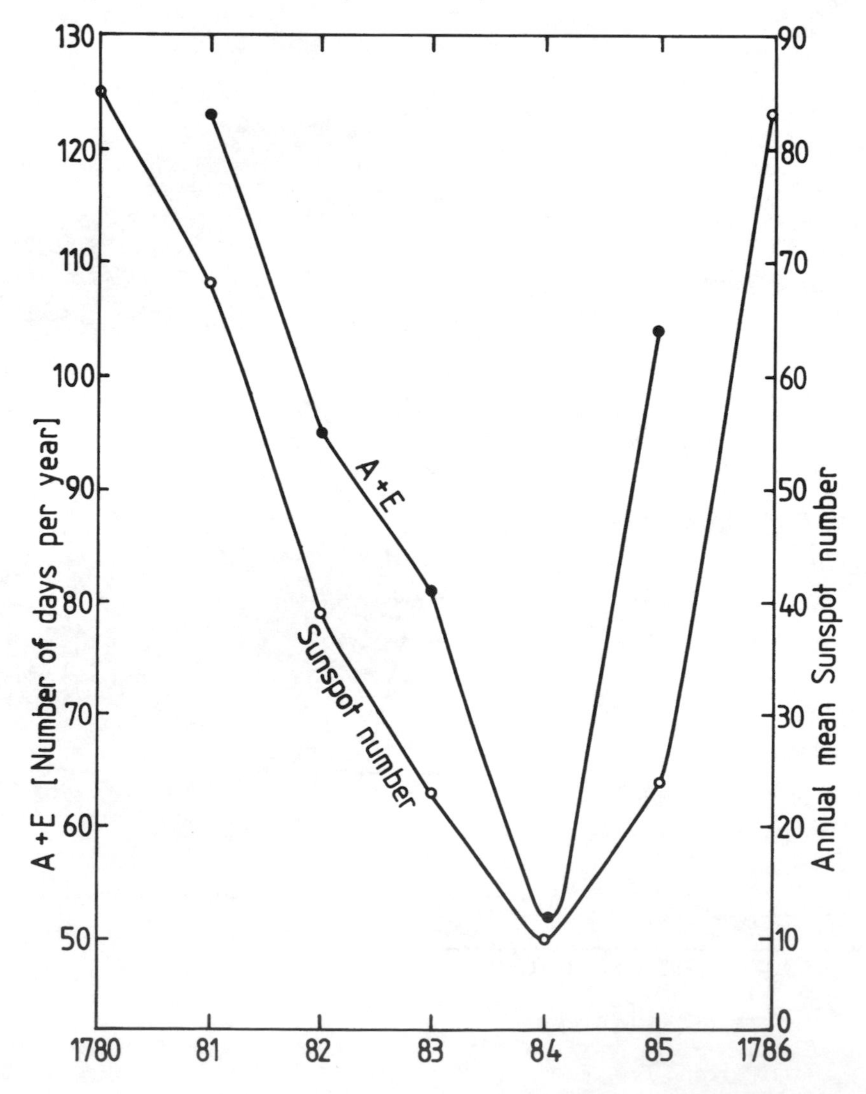

Table 6.15 *Catalogue of* Grosswetterlagen *for 1781.*

	Jan	Feb	Mar	Apr	May	Jun	Jul	Aug	Sep	Oct	Nov	Dec
1	SWA	WZ	BM	HFA	HNFZ	HM	HM	BM	HM	BM	BM	HFA
2	SWZ	WZ	BM	HFA	HNFZ	HNZ	HM	BM	HM	U	BM	HFA
3	NZ	WA	BM	HFZ	HNZ	HNZ	HM	U	HM	BM	BM	HFA
4	NZ	WZ	BM	HFA	HNZ	HNZ	BM	BM	HM	BM	BM	HFA
5	BM	BM	BM	SEZ	HNZ	TB	BM	BM	BM	BM	SWZ	HFA
6	BM	BM	WA	SEZ	HNZ	TB	BM	HFZ	BM	HB	SWZ	HNFA
7	BM	BM	WA	SWZ	HNA	TB	TB	HFZ	BM	HB	SWZ	HNFA
8	BM	SWA	WA	WZ	HNA	TB	TB	HFZ	BM	HB	BM	HNFA
9	BM	SWA	WA	WZ	HFZ	TB	WZ	HFA	BM	HM	BM	SEA
10	HNFA	SWA	HB	WA	HFZ	TB	WZ	HFA	BM	HM	BM	SEA
11	HNFA	SWA	HB	WA	HFZ	HFZ	WZ	BM	HFA	HM	WZ	SEA
12	HNFA	SWA	HB	WA	HM	HFZ	BM	BM	HFA	BM	WZ	SA
13	SEA	WZ	HB	WA	HM	HFZ	BM	BM	HFA	BM	WZ	SA
14	SEA	WZ	HB	WA	TB	HFZ	BM	BM	HM	BM	WZ	SA
15	SEA	WZ	HB	WZ	TB	HFZ	BM	TB	HM	BM	WZ	SA
16	SA	WZ	HB	HM	TB	HFZ	HB	TB	SWZ	BM	WZ	U
17	SA	WZ	HB	HM	HNFA	HFA	HB	TB	WZ	BM	WZ	SZ
18	TRW	WZ	BM	HM	HNFA	HFA	NZ	TRW	WZ	BM	WZ	SA
19	TRW	WZ	BM	BM	HNFA	HFA	NZ	TRW	WA	NWZ	BM	SA
20	TRW	NEZ	BM	BM	U	HNA	NZ	TM	WA	NWZ	BM	SA
21	TB	NEZ	BM	BM	HB	HNA	NZ	TM	WZ	NWZ	BM	HM
22	TB	NEZ	NWA	BM	HB	HNA	NZ	U	WZ	NZ	SWA	HM
23	TB	NWZ	NWA	BM	HB	HNA	NWZ	HM	TRM	NZ	SWA	SWA
24	TB	NWZ	NWA	U	HB	HNZ	NWZ	HM	TRM	BM	SA	SWA
25	TB	TRW	NWA	U	HB	HNZ	BM	WZ	TRM	U	BM	SWA
26	TB	TRW	NZ	NEA	HB	HNZ	BM	WZ	TM	BM	BM	SWA
27	WZ	TRW	NZ	NEA	HB	BM	BM	WA	TM	BM	HFA	WZ
28	WA	NWZ	NA	HFZ	HB	BM	BM	WA	TM	BM	HFA	WZ
29	WA		NA	HFZ	HM	BM	HM	WA	BM	TRM	HFA	WZ
30	WA		NA	HFZ	HM	BM	HM	HM	BM	U	SA	WZ
31	WZ		NA		HM		HM	HM		TRM		WZ

Table 6.16 *Catalogue of* Grosswetterlagen *for 1782.*

	Jan	Feb	Mar	Apr	May	Jun	Jul	Aug	Sep	Oct	Nov	Dec
1	U	BM	BM	TB	HNZ	WZ	WA	WZ	BM	BM	WZ	HFA
2	WZ	BM	BM	TB	HNA	WZ	WA	WZ	BM	TB	WZ	HFA
3	WZ	WS	BM	TB	HNA	WZ	WA	WZ	HM	TM	WZ	HFA
4	WZ	WS	BM	TB	HNZ	WZ	WA	WZ	HM	TM	WZ	HFA
5	WZ	WS	WA	TB	TM	HB	WA	WZ	HFA	TM	WZ	HFA
6	WZ	WS	WA	TB	TM	HB	WS	WZ	HFA	TM	TRM	HFA
7	WZ	WS	WA	U	TM	HM	WS	WZ	HFA	HFZ	TRM	HFA
8	WZ	HFZ	WZ	HFZ	HFZ	HM	WS	WZ	HFA	HFZ	TRM	HFA
9	WZ	HFZ	WZ	HFZ	HFZ	TB	WS	WZ	HFA	HFA	HFZ	HFA
10	WZ	HNZ	WZ	HFZ	HFZ	TB	WS	WZ	HFA	HFA	HFZ	HFA
11	BM	HNZ	WZ	TB	TB	TB	BM	WZ	HNFA	U	HFA	HFA
12	BM	HNZ	WZ	TB	TB	WA	BM	WZ	HNFA	HNZ	HFA	BM
13	BM	HNZ	NWZ	TB	WZ	WA	HM	WZ	HNFA	HNZ	BM	BM
14	BM	HB	NWZ	TB	TB	HM	HM	WZ	U	BM	BM	BM
15	BM	HB	HNA	HNZ	TB	HM	HM	WZ	TB	BM	WZ	U
16	WZ	HB	HNA	HNZ	TB	HM	HNA	WZ	TB	BM	WZ	BM
17	WZ	HB	HNZ	HNZ	TB	HFA	HNZ	WZ	WZ	BM	NZ	BM
18	WZ	HB	U	HNFA	WZ	HFA	HNA	WZ	WZ	WA	NZ	BM
19	WA	HB	WZ	HNFA	WZ	BM	HM	WA	WZ	WZ	NZ	BM
20	WA	HM	WZ	HNFA	U	BM	HM	WA	WA	WZ	NZ	BM
21	WA	HM	WS	TB	TB	HB	HM	WA	WA	WZ	NEZ	NWZ
22	WA	SWA	WS	TB	TB	HB	HM	WZ	U	WA	NEZ	NWZ
23	WZ	SWA	WS	TB	TB	HB	HM	WZ	WZ	WA	HFZ	NWZ
24	WZ	WA	WS	TRW	TB	HB	HM	BM	WZ	BM	HFZ	NWZ
25	WZ	WA	U	TRW	WZ	HFA	HM	BM	WA	BM	HFZ	NWZ
26	WS	WA	BM	HNFA	WZ	HFA	U	BM	WA	BM	HFZ	NWZ
27	WS	WA	BM	HNFA	HM	BM	TB	BM	BM	BM	HFZ	WA
28	WS	BM	SWA	HNFA	HM	BM	TB	BM	BM	BM	HFZ	WA
29	WS		SWA	HNFA	TB	BM	U	BM	BM	WZ	HFZ	WA
30	WS		SWZ	HNA	TB	BM	WA	BM	BM	WZ	HFA	NA
31	WS		TB		WZ		WA	BM		WZ		NA

Table 6.17 *Catalogue of* Grosswetterlagen *for 1783.*

	Jan	Feb	Mar	Apr	May	Jun	Jul	Aug	Sep	Oct	Nov	Dec
1	NZ	BM	TRM	BM	HNFA	HNFZ	HM	HM	BM	HM	HM	HM
2	NWA	BM	TRM	BM	HNZ	HNFA	HM	HM	BM	BM	HM	HM
3	NWA	BM	U	BM	HNZ	HNFA	BM	HM	BM	BM	HM	SA
4	SWA	BM	U	BM	BM	HFA	BM	U	WZ	WA	HFA	SA
5	WZ	WZ	TB	NA	BM	HFZ	BM	HM	WZ	WA	HFA	SA
6	WZ	WZ	TB	NA	HNZ	HFZ	WA	HM	WZ	SWA	HFA	HFA
7	WZ	WZ	WZ	NWA	HNZ	BM	WA	WZ	WZ	SWA	HNFA	HFA
8	WZ	WZ	WZ	NWA	BM	BM	HM	WA	WZ	WZ	HNFA	HFA
9	WW	WZ	WZ	NWA	BM	BM	HFA	WA	WZ	BM	HNFZ	HM
10	WW	WZ	TRW	HM	BM	BM	HFA	WA	WZ	BM	HNZ	HM
11	WZ	WA	TRW	HFA	HFA	WZ	HFA	TRM	WA	HM	HNZ	HM
12	WZ	WA	HNZ	HFA	HFA	WZ	HFA	TRM	WA	HM	WZ	HM
13	WZ	WA	HNZ	BM	HFA	TB	HFA	TRM	WA	HFA	WZ	HM
14	WZ	U	HNZ	BM	HFA	TB	HFA	TRM	WA	HFA	WZ	HM
15	WS	HB	HNFA	BM	BM	TB	HFA	BM	TRW	BM	WZ	BM
16	WS	HB	HNFA	NWZ	BM	TB	HFA	BM	TRW	BM	WW	BM
17	WS	HB	NWA	NWA	BM	WA	BM	HFA	BM	HM	WW	BM
18	WS	BM	NWA	NWZ	HNZ	WA	BM	HFA	BM	HM	WZ	BM
19	WS	BM	NWA	BM	HNZ	TRW	BM	HFA	SWA	HM	WZ	BM
20	U	BM	HM	BM	HNFA	TRW	BM	HFA	SWA	HM	WZ	BM
21	TB	WZ	HM	TRM	HNFA	TRM	WZ	BM	SWA	HFA	WZ	HM
22	TB	WZ	HM	TRM	HFA	TRM	WA	BM	BM	HFA	BM	HM
23	TB	WZ	WZ	HNZ	HNZ	BM	WZ	BM	BM	WZ	BM	HFZ
24	SWZ	WZ	WZ	HNZ	HNZ	BM	BM	TB	BM	WA	NWZ	HFZ
25	SWZ	TRM	WZ	HNFZ	U	BM	BM	TB	HM	WA	NWZ	TB
26	SWZ	TRM	WZ	HNFA	TM	BM	HM	TB	HM	U	BM	TB
27	SWZ	TRM	WZ	HFA	TM	BM	HFA	HM	BM	WZ	BM	WS
28	SWZ	TRM	NZ	HFA	TM	BM	HFA	HM	BM	WZ	HM	WS
29	WZ		NZ	HFA	TB	BM	BM	BM	BM	WA	HM	U
30	WZ		BM	HFA	TB	HM	BM	BM	HM	WA	HM	U
31	WZ		BM		HFZ		BM	BM		HM		SEA

Table 6.18 *Catalogue of* Grosswetterlagen *for 1784.*

	Jan	Feb	Mar	Apr	May	Jun	Jul	Aug	Sep	Oct	Nov	Dec
1	SEA	NWZ	BM	BM	TM	HFA	NWA	BM	U	HNFA	TB	BM
2	SEA	NWZ	BM	BM	TM	HNFA	NWA	BM	BM	HFA	TB	BM
3	HFA	NWZ	BM	BM	TM	HNFA	NWA	BM	HM	HFA	HM	WZ
4	HFA	WZ	SWA	BM	NWA	HNFA	BM	BM	HM	HFA	HM	WZ
5	HFA	WZ	SWA	BM	NWA	U	BM	BM	HM	HFA	HNFA	WZ
6	HFA	TRW	SWA	BM	NWA	U	BM	WA	HM	HFA	HNFA	TB
7	HFA	TRW	TB	BM	WA	U	TRW	WZ	HM	HFA	BM	TB
8	HFA	TRW	TB	BM	WA	U	TRW	WZ	HM	HFA	BM	TB
9	BM	TRM	TB	U	WA	WZ	BM	BM	HM	HFA	WA	TB
10	BM	TRM	WZ	U	U	WZ	BM	BM	NA	HNFZ	WA	WS
11	BM	TRM	WZ	TB	TRM	WZ	BM	BM	HM	HNFZ	WZ	WS
12	BM	HNFZ	WZ	TB	TRM	WZ	BM	BM	HM	HNFZ	WZ	TM
13	BM	HNFZ	WZ	TB	BM	WZ	BM	BM	HM	HM	WZ	TM
14	BM	HNFZ	BM	TB	BM	WZ	BM	BM	HM	HM	WA	TM
15	WZ	HNFZ	BM	TB	BM	WA	BM	BM	HM	HM	WA	TM
16	WZ	HNFZ	BM	U	BM	WA	BM	BM	HM	BM	WA	TM
17	WZ	HNFZ	BM	U	BM	WA	WZ	U	HM	BM	WA	HNFZ
18	TM	HFA	WS	BM	BM	WA	WA	TM	HM	BM	WZ	HNFZ
19	TM	HFA	WS	BM	BM	BM	WZ	TM	TRW	WA	WZ	TRM
20	TM	HFA	WS	WZ	BM	BM	WZ	U	TRW	WA	WZ	TRM
21	TM	HFA	BM	WZ	BM	TB	WZ	TB	WZ	WA	NZ	TRM
22	TRM	HFA	BM	WZ	BM	TB	WZ	TB	WA	BM	NZ	TRM
23	TRM	WW	BM	WZ	BM	TB	WZ	TB	WZ	WZ	WZ	TRM
24	TRM	WW	TB	WZ	BM	WZ	WZ	TB	HM	WZ	WA	HB
25	BM	WW	TB	WZ	TRW	WZ	BM	TB	HM	WZ	WA	HB
26	BM	WW	TB	BM	TRW	WZ	BM	WZ	TB	BM	BM	HB
27	BM	WW	WS	BM	TRW	WZ	WZ	WZ	TB	BM	BM	HM
28	HNFZ	BM	WS	U	BM	WZ	WA	WZ	TM	BM	BM	HM
29	HNFZ	BM	WS	BM	BM	WZ	TB	WZ	TM	TM	BM	HM
30	NZ		WS	BM	HFA	WZ	TB	WZ	HNFZ	TM	BM	TRW
31	NZ		WS		HFA		BM	WZ		TB		TRW

Table 6.19 *Catalogue of* Grosswetterlagen *for 1785.*

	Jan	Feb	Mar	Apr	May	Jun	Jul	Aug	Sep	Oct	Nov	Dec
1	TRW	TM	HNA	TRW	HNFA	WZ	U	WZ	TB	NWA	TB	WZ
2	TB	HFZ	HNA	TRW	U	WZ	WZ	WZ	TB	NWA	TB	SWA
3	TB	HFZ	HNA	NZ	BM	WZ	WZ	TB	TB	SA	TB	SWZ
4	TB	U	HNA	NZ	BM	TB	WZ	TB	TB	SA	TB	SWZ
5	TB	TB	HNA	NEZ	BM	TB	NWZ	TB	TB	SA	TB	WZ
6	SWZ	TB	HNA	NEZ	BM	TM	NWZ	WZ	TB	SWA	WZ	SA
7	BM	WZ	HNA	NEZ	BM	TM	NWZ	WZ	TB	WA	WZ	SA
8	BM	WZ	TRW	NEZ	BM	BM	NWZ	WZ	U	WA	WZ	SA
9	BM	WZ	TRW	NEA	BM	BM	NWZ	WZ	U	WA	BM	SEA
10	BM	WZ	TRW	NZ	HNA	BM	NWZ	WZ	TB	WA	BM	HFA
11	SA	HNA	U	BM	HNA	BM	WZ	WZ	TB	WA	BM	HFA
12	SA	HNA	HNZ	BM	HNA	BM	WZ	TB	U	WA	WZ	SEA
13	SA	HNA	HNA	BM	BM	NWA	WZ	TB	U	WA	WZ	SEA
14	SA	HNA	HNA	BM	BM	U	TB	WZ	HM	WA	BM	SA
15	SA	NZ	HNA	BM	NWZ	U	TB	WZ	HM	WA	BM	SA
16	SA	NZ	HNA	BM	NWZ	TM	WZ	WZ	WZ	WA	BM	SEA
17	SWA	TM	NA	SA	WZ	TM	WZ	WZ	WZ	NWA	BM	SEA
18	SWA	TM	NZ	SA	WZ	TM	WZ	WZ	WZ	NWA	BM	SEA
19	SWA	TB	NZ	SA	WZ	NEZ	WZ	WZ	U	NWA	TRW	SEA
20	SA	TB	NZ	WA	WZ	NEZ	WZ	WZ	WA	NA	TRW	SEA
21	SA	TB	HNZ	WA	WZ	NZ	WZ	WZ	WZ	NA	TRW	SEA
22	SA	TB	HNA	WA	HM	NZ	WZ	WZ	WZ	BM	BM	HFZ
23	SA	TM	HNA	U	HM	NZ	WZ	WZ	WA	BM	BM	HFZ
24	SA	U	HNZ	BM	HM	BM	WA	WA	WZ	WA	BM	HFZ
25	SA	TM	HB	NA	BM	BM	HM	WZ	WZ	WA	BM	HFZ
26	SA	HFZ	HB	NA	BM	BM	HM	WZ	WZ	WA	WZ	BM
27	SWA	HFZ	TRM	NA	WZ	BM	HM	WA	NWZ	WA	WZ	BM
28	SWA	HNA	TRM	NA	WZ	HNA	TB	WA	NWZ	WA	WZ	HNZ
29	WZ		TRM	HNFA	WZ	HNA	TB	WA	NWA	WA	WZ	HNZ
30	WZ		TRM	HNFA	NWZ	HNA	TB	WA	U	WA	WZ	HNZ
31	TM		TRM		NWZ		TB	WZ		WA		HNA

Table 6.20 Grosswetterlagen: *annual and period average frequencies of the three main circulation patterns; 1781–5 and 1881–1980.*

	Number of days per year		
	Zonal	Mixed	Meridional
1781	55	145	156
1782	123	97	134
1783	85	154	117
1784	96	150	107
1785	107	91	153
Period averages			
1781–5	93.2	127.4	133.4
1881–1980	97.8	116.4	148.4

by the weather. Following a series of good years for harvests 1781 was a turning point for both weather and agriculture; a number of unfavourable events such as spring and summer droughts, late springs, severe winters and an unusual dust haze, all played their part in affecting the affairs of man during the 1780s. The most crucial event of all was the long and severe winter of 1788–9 in France, which, following the disastrous harvest of 1788, resulted in food prices rising sharply and mass unemployment, with thousands of people flocking to Paris from the impoverished countryside (Rudé, 1965). The pre-revolutionary economic and social discontent was certainly intensified by the harsh weather conditions. As synoptic charts become available for the years 1788 and 1789 it will be interesting to investigate the daily sequence of circulation patterns that gave rise to this extremely cold winter.

Bibliography and references

Baur, F., *Musterbeispiele Europäischer Grosswetterlagen*, Dieterich'sche Verlagsbuchhandlung, Wiesbaden, 1947.

Brezowsky, H., Flohn, H. and Hess, P., Some remarks on the climatology of blocking action, *Tellus*, **3** (1951), pp. 191–4.

Desaive, J.-P., Goubert, J.-P., Le Roy Ladurie, E., Meyer, J., Muller, O. and Peter, J.-P., *Médecins, Climat et Epidémies à la Fin du XVIIIe Siècle*, Mouton & Co., Paris & The Hague and Ecole Pratique des Hautes Etudes, Paris, 1972.

Douglas, K. S., Lamb, H. H. and Loader, C., *A Meteorological Study of July to October 1588: the Spanish Armada Storms*, CRU RP6, Climatic Research Unit, School of Environmental Sciences, University of East Anglia, Norwich, 1978.

Douglas, K. S. and Lamb, H. H., *Weather Observations and a Tentative Meteorological Analysis of the Period May to July 1588*, CRU RP6a, Climatic Research Unit, School of Environmental Sciences, University of East Anglia, Norwich, 1979.

Hess, P. and Brezowsky, H., *Katalog der Grosswetterlagen Europas (1881–1976)*, Berichte des Deutschen Wetterdienstes, Nr. 113, Offenbach am Main, 1977.

Kington, J. A., A late eighteenth century source of meteorological data, *Weather*, **25** (1970), pp. 169–75.

Kington, J. A., Daily weather mapping from 1781: a detailed synoptic examination of weather and climate during the decade leading up to the French Revolution, *Climatic Change*, **3** (1980), pp. 7–36.

Kington, J. A., Meteorological observing in Scandinavia and Iceland during the eighteenth century, *Weather*, **27** (1972), pp. 222–33.

Kington, J. A., The *Societas Meteorologica Palatina:* an eighteenth century meteorological society, *Weather*, **29** (1974), pp. 416–26.

Lamb, H. H., Volcanic dust in the atmosphere; with a chronology and assessment of its meteorological significance, *Philosophical Transactions of the Royal Society*, **A266** (1970), pp. 425–533.

Lamb, H. H., British Isles Weather Types and a Register of the Daily Sequence of Circulation Patterns, 1861–1971, *Geophysical Memoir*, **116**, HMSO, London, 1972.

Lamb, H. H. and Johnson, A. I., Secular Variations of the Atmospheric Circulation since 1750, *Geophysical Memoir*, **110**, HMSO, London, 1966.

Ludlam, F. H., *The Cyclone Problem: a History of Models of the Cyclonic Storm*, Imperial College of Science and Technology, London, 1966.

Murray, R. and Lewis, R. P. W., Some aspects of the synoptic climatology of the British Isles as measured by simple indices, *Meteorological Magazine*, **95** (1966), pp. 193–203.

Oliver, J. and Kington, J. A., The usefulness of ships' log-books in the synoptic analysis of past climates, *Weather*, **25** (1970), pp. 520–8.

Rudé, G. E., The outbreak of the French Revolution, *The American and French Revolutions, 1763–93*, (Ed. A. Goodwin), *The New Cambridge Modern History*, Volume VIII, Cambridge University Press, London, 1965.

Schneider-Carius, K., *Wetterkunde Wetterforschung*, Karl Alber, Freiburg, 1955.

Shaw, Sir Napier, *The Drama of Weather*, Cambridge University Press, London, 1933.

See also

Climate Monitor, 1–16 (1972–86), Climatic Research Unit, University of East Anglia, Norwich.

Die Grosswetterlagen Europas, **30–9** (1977–86), Deutscher Wetterdienst, Offenbach am Main.

Index

Académie de Médecine, 11
Accademia del Cimento (Academy of Experiments), 4
Air masses, 20, 21, 26
America, North, 12, 18

Bacon, Francis, 3
Balchin, W. G. V., 2
Barker, Thomas, 14, 25
Baur, F., 148, 165
Beaufort, Francis (Admiral), 12
Beaufort scale, 12, 16
Beiträge zur Witterungskunde (H. Brandes), 19
Bergen, School of Meteorology, 20
Bering, V., 5
Bjerknes, Vilhelm and Jacob, 20
Blocking, 149–52, 155, 156
Boerhaave, Hermann, 4, 5
Boston, USA, 18
Bracknell, Meteorological Office, 2, 25
Brandes, Heinrich, 18, 19, 20
Brezowsky, H., 148, 153, 156, 165
British Isles, 14, 20, 149, 150, 153
Buys Ballot, Christophorus, 19

Carbon dioxide, 2
Circulation patterns (*See also* Lamb British Isles Weather Types and *Grosswetterlagen*), 2, 3, 22, 148–64
Clermont-Ferrand, 3
Climate Monitor, 149, 165
Climatic change, 1, 2, 149
Climatic Research Unit, 2
Cockatrice (HM Cutter), 26
Cotte, Louis (*Le Père*), 5

Daily weather maps, 1, 2, 3, 18, 19, 20, 21, 26, 27–147, 153, 158, 164
d'Azyr, Vicq, 5
Defoe, Daniel, 18
Delisle, temperature scale, 22
Desaive, J.-P. *et al.*, 11, 165
Deutscher Wetterdienst, 153
Douglas, K. S. *et al.*, 1, 165

Electromagnetic telegraph, 19
Enlightenment, 11, 13, 15
Ephemerides (Societas Meteorologica Palatina), 12, 14, 22
'European Monsoon', 152

Ferdinand II, Grand Duke of Tuscany, 3
Fitzroy, Robert (Admiral), 20
Flohn, H., 156
Florence, 4
France, 5, 164
Franklin, Benjamin, 18
French Revolution, 11, 14, 158
Fronts, 20, 26

Galton, Francis, 19, 20
Geostrophic wind, 21, 26
Germany, 5, 11
Goethe, J. W. von, 5
Great Northern Expedition to Siberia, 5
Greenland, 12
Grosswetterlagen, 148, 153–6, 158–63
Grosswetterlagen Mitteleuropas, Die, 153

Hemmer, Johann, 12, 14, 22
Hess, P., 148, 153, 156, 165
Hooke, Robert, 4, 14
Howard, Luke, 14

Iceland, 14, 15
Italy, 4

Japan, 14
Johnson, A. I., 2, 165
Jurin, James, 4, 5, 12, 14

Karl Theodor, Prince-Elector of the Palatinate, 11
Kington, John, 5, 12, 15, 158, 165

Ladurie, Emmanuel Le Roy, 11, 165
Lamb British Isles Weather Types, 148–58
Lamb, Hubert, 2, 148, 149, 165
Lambhús, Iceland, 15, 25
Lavoisier, Antoine, 5, 11
Lewis, R. P. W., 151, 165
Lievog, Rasmús, 15, 25
Little Ice Age, 2, 153
Logbooks, Naval, 15–16
London (Royal Society), 4, 5, 14
Louis XVI, 5
Lyndon Hall, Rutland, 14, 25

Mannheim, 11, 12, 14, 18, 22
Maret, Dr, 24, 25
Meteorographica or Methods of Mapping the Weather (F. Galton), 20
Meteorological Office, UK, 2, 25 *passim*
Meteorological Research Committee, 2
Montmorency, 5
Murray, R., 150, 165
Musschenbroek, Pieter van, 12

National Maritime Museum, 15
Norway, 20
Norwegian Weather Service, 20
Norwich, 2, 26

Oliver, John, 2, 15, 165

Palatinate of the Rhine, 11, 12
Paris, 3, 5, 11, 18, 25, 164
Paris inches and lines, 5, 12
Philadelphia, 18
Philosophical Transactions of the Royal Society, 5
Prague, 12
PSCM indices, 151–3
Public Record Office, 15

Réaumur, temperature scale, 5, 12, 22
Royal Society, 4, 5, 12, 14
Rudé, G. R., 164, 165

Sachsen-Weimar, 5
Schneider-Carius, K., 19, 165
Siberia, 5, 12
Societas Meteorologica Palatina, 11–14, 18, 19, 22, 25
Société Royale de Médecine, 5, 11, 12, 24, 25
Sorbonne, University of Paris, 11
Spanish Armada, 1
Stevenson screen, 12
Stockholm, 3
Strnadt, Antony, 12
Sunspot numbers, 156
Sun–weather relationship, 156
Swansea, 2
Synoptic climatology, 3, 148, 151
Synoptic meteorology, 3, 18–20
Synoptic weather charts/maps (*See* daily weather maps)

Traité de Météorologie (L. Cotte), 5

University College of Swansea, 2
University of East Anglia, Norwich, 2
University of Leyden, 4

Volcanic activity, 2, 14

'Weather Journals', 4
Weather lore, 3
Weather types (*See also* Lamb British Isles Weather Types and *Grosswetterlagen*), 3, 21, 148–64
Westerly wind, westerly weather type etc., 149–51

Youell, William, 26